Series in BioEngineering

The Series in Bioengineering serves as an information source for a professional audience in science and technology as well as for advanced students. It covers all applications of the physical sciences and technology to medicine and the life sciences. Its scope ranges from bioengineering, biomedical and clinical engineering to biophysics, biomechanics, biomaterials, and bioinformatics.

Indexed by WTI Frankfurt eG, zbMATH.

More information about this series at http://www.springer.com/series/10358

Meir Israelowitz · Birgit Weyand ·
Herbert P. von Schroeder · Peter Vogt ·
Matthias Reuter · Kerstin Reimers
Editors

Biomimetics and Bionic Applications with Clinical Applications

Springer

Editors
Meir Israelowitz
Biomimetics Technologies Inc.
Toronto, ON, Canada

Herbert P. von Schroeder
University of Toronto
Toronto, ON, Canada

Matthias Reuter
Technical University
Clausthal-Zellerfeld
Niedersachsen, Germany

Birgit Weyand
Abteilung für Plastische und Handchirurgie
Medizinische Hochschule Hannover
Hannover, Germany

Peter Vogt
Medizinische Hochschule Hannover
Hannover, Niedersachsen, Germany

Kerstin Reimers
(Deceased)
Hannover, Germany

ISSN 2196-8861 ISSN 2196-887X (electronic)
Series in BioEngineering
ISBN 978-3-319-53212-7 ISBN 978-3-319-53214-1 (eBook)
https://doi.org/10.1007/978-3-319-53214-1

This Springer imprint is published by the registered company Springer Nature Switzerland AG
The registered company address is: Gewerbestrasse 11, 6330 Cham, Switzerland

This volume is dedicated to Prof. Kerstin Reimers and the Fadhlaoui Family. During the completion of this work, Prof. Reimers our co-editor sadly succumbed to a terminal illness.

Preface

This volume focuses on biomimetic application to the clinical level, while biomimetic itself, mimic nature can be done in twofolds, one external refereeing to the bionic aspects and molecular level interaction the biomimetic. For any technology to have any clinical impact need not only to overcome the technological issues, but regulatory.

Our intent in the volume is not only to create a theoretical foundation but also show direct examples in nature, where can be transformed into therapeutic solutions, but technologies to rise to this level. We are not only showing physical examples, but also biomimetic methods can be implemented in the clinical level.

Accordingly, the chapters follow in fourth categories to underline the biomimetic solutions in the clinical level, the first chapters concentrated on theoretical foundation, challenge, and biomimetic in the surgery level. The second section concentrated specific examples from the bacteria, to especially sensors, into higher biological systems. The third section concentrated on technology which works under biomimetic principals. The last is methods, implement through software implementing biomimetic principles.

The contributors of this volume represent three continents, especially places with a challenge to developed new technologies limited by the resources in the community or the misuse of resources, but because there is not a specific place where biomimetic is specific, the possibility of any location can be developed, since biological systems through evolution had taken place, even in the most austere and challenge place, because this development of biomimetic can be found any location where nature took place. The chapters cover issues from bacterial sensors and complex systems for the development of cells to more advanced therapeutic applications in the clinical level. The software considers as an example is neural networks as a method to mimic to solve clinical and developed therapeutic solutions.

The volume not only demonstrated developments in biomimetic to clinical work but also can be applied to other fields from the biomedical sciences, but everyday technologies, as magnetic detectors, far-infrared detectors, material development, material grow, and software development.

Toronto, Canada — Meir Israelowitz
Hannover, Germany — Birgit Weyand
Toronto, Canada — Herbert P. von Schroeder
Hannover, Germany — Peter Vogt
Clausthal-Zellerfeld, Germany — Matthias Reuter
Hannover, Germany — Kerstin Reimers[1]

April 2017

[1]Late Prof. Reimers succumbed to terminal illness in the process working in the volume.

Contents

Biomimetics Theoretical Foundation

Biomimetics: A Biosemiotic View 3
Kalevi Kull

Medical Biotechnology and Biomimetics: Prospects and Challenges in Sub-Saharan Africa 19
Obaro S. Michael

Biomimetics and Its Influence in Plastic and Reconstructive Surgery 29
Birgit Weyand and Peter Vogt

Biomimetics Models

Torsional Magnetic Angle for *Magnetospirillum gryphiswaldense* 47
Sarah Strauß, Meir Israelowitz, Birgit Weyand, Robert Müller, Henkel Thomas, Dirk Schüler, René Uebe, Syed W. H. Rizvi, Christoph Gille, Herbert P. von Schroeder, Kerstin Reimers, and Peter Vogt

Spider Silk as Biomaterial for Medical Applications and Tissue Engineering 61
Malte Fließ and Sarah Strauß

The Innovative Power of the Electric Eel (*Electrophorus electricus*) 71
Jenifer Gifford and Matthew Leming

Morphological Study of the Infrared Sensory Pits of Pit Viper, Python and Boa Snakes 81
Birgit Weyand, Meir Israelowitz, Matthias Reuter, Sabine Bohlmann, Robert Wagner, Syed W. H. Rizvi, Chistoph Gille, Kerstin Reimers, Peter Vogt, and Herbert P. von Schroeder

Biomimetics Applications

Optical Oxygen Measurements Within Cell Tissue Using Phosphorescent Microbeads and a Laser for Excitation 107
Elmar Schmälzlin, Mariel Nöhre, and Birgit Weyand

Emerging Biomimetic Approaches in the Optimization of Drug Therapies 131
Obaro S. Michael

Biomimetics Strategies to Overcoming Noise 147
Syed W. H. Rizvi, Birgit Weyand, Meir Israelowitz, Christoph Gille, Matthias Reuter, Sabine Bohlmann, Kerstin Reimers, Peter Vogt, and Herbert P. von Schroeder

Computer Models in Biomimetics

Biological Inspired Optical Pattern Analysis by Topological Neurons 159
Matthias Reuter and Sabine Bohlmann

***Neural Networks* for Modeling Metabolic Pathways** 177
Meir Israelowitz, Birgit Weyand, Sabine Bohlmann, James Kramer, Christoph Gille, Syed W. H. Rizvi, Herbert P. von Schroeder, and Matthias Reuter

Computer-Based Intelligence Methods Applied for Personalized Management of Diabetes 195
Matthias Reuter and Sabine Bohlmann

Biological Inspired Image Analysis for Medical Applications 211
Matthias Reuter and Sabine Bohlmann

Author Index 227

Biomimetics Theoretical Foundation

Biomimetics: A Biosemiotic View

Kalevi Kull

Abstract After giving a brief review of different approaches related to biomimetics, we focus on the question of the general mechanisms of problem-solving and self-building in living systems. It is possible to develop a theory of biomimetics in connection with a semiotic approach to understand the workings of living systems. In this case we concentrate on the innovative or knowledge-acquisition aspects and mechanisms of life. This would make it possible to understand why living systems are suitable and specific to be used as models for technological modelling.

Keywords Biosemiotic technology · Biosemiotics · Ecosemiotics · History of biology · Sustainable technology · Theory of knowledge

Biomimetics is a smart technology that is based on the smartness of life itself. In other words, biomimetics is a technological approach that uses the knowledge acquired by living systems. This is an approach to apply the building principles of non-human life in the design of various technical systems in human culture. Biomimetics uses and remakes what is meaningful (which includes functional[1]) in life's findings.

Biomimetics is based on understanding that life has been able to construct what non-life cannot construct; that life is capable for making discoveries that are hardly accessible otherwise; that living systems, from a living cell to coenoses are capable to acquire and carry knowledge. Everything that humans construct requires the use of knowledge. Likewise, the materials used by organisms and the structures they build, require finding and recognizing substances, and remembering the steps of construction. This is what differentiates life's production from non-mediated chemistry.

[1]On the relationship between meaning and function, see [11].

K. Kull (✉)
Department of Semiotics, University of Tartu, Jakobi St. 2, 51005 Tartu, Estonia
e-mail: kalevi.kull@ut.ee

M. Israelowitz et al. (eds.), *Biomimetics and Bionic Applications with Clinical Applications*, Series in BioEngineering,
https://doi.org/10.1007/978-3-319-53214-1_1

This approach assumes that living systems have certain knowledge, which also means that living systems are capable of acquiring new knowledge. This idea is not new, however, the understanding of these mechanisms and acknowledging its central importance for biology has still only received little attention.

Part of the problem is that the concept of knowledge needs to be defined more clearly. For instance, are the adaptations of fireflies or *Escherichia coli* considered to be knowledge-based? However, if we state that knowledge is just any kind of meaningful information, then the applicability of the concept throughout living systems is feasible.

Both J. B. Lamarck and Charles Darwin can be credited as providing explanations of life's capacity for innovations, or in their terms, as the formation of adaptations. The direct connection of this problem to epistemology, i.e., gaining knowledge, is more the work of the twentieth century: first by J. M. Baldwin (organic selection), then by J. Piaget (genetic epistemology), and H. Maturana (biology of knowledge).

A catalytic role in theoretical biology of the twentieth century has been played by the Theoretical Biology Club (founded in 1932 in England), where J. Woodger (with his interest in logic of biology), C. H. Waddington (studying epigenetics), L. L. Whyte (who introduced the concept of internal factors of evolution), and several others discussed the fundamentals of life [1, 44]. Karl Popper, J. Woodger's friend, also attended this club. In his lecture of 1989, "Towards an evolutionary theory of knowledge", Popper made several remarkable points: "[…] in the biological and evolutionary sense in which I speak of knowledge, not only animals and men have expectations and therefore (unconscious) knowledge, but […], indeed, all organisms. […] All adaptations to environmental and to internal regularities, to long-term situations and to short-term situations, are kinds of knowledge […]. Thus, the origin and the evolution of knowledge may be said to coincide with the origin and the evolution of life" [52: 61, 64]. And he adds that "adapted forms will be some of those forms which responded to a challenge, which solved problems" [53: 133].

Knowledge in any occasion assumes and requires sign relations and meaning-making. Sign relations means that something is *about* something. Sign processes comprise the study area for *semiotics*. Thus the science whose object of study is knowledge is semiotics, while the prelinguistic (i.e., biological) forms of knowledge are the objects of biosemiotics [36].

Biosemiotics as a contemporary approach in biology is a study of life's meaningfulness, or rather its processes of meaning-making. Biosemiotics is the theory of how meaningful life works, how it develops its findings, how it makes its searches, how it makes and carries meanings.[2] Biomimetics is a technological approach to make use of understanding of how life does this by mimicking or imitating, or just by being inspired by life's discoveries.[3]

[2]For a contemporary overview on biosemiotics, see [12, 14, 26].

[3]See the recent reviews of biomimetics, e.g., [2, 5]. Couple of volumes on biomimetics were recently published by Springer: [23, 45, 51] (see also [10]).

Biomimetics also has an epistemological role in developing an understanding of life.[4] By mimicking the mechanisms of organisms it is possible to come closer to the enigma of life. Since the natural epistemology is one of major focuses of biosemiotics, we could say that *biomimetics is a technological biosemiotics*. This aspect is also pointed out by Webb [64] and Rossi and Pieroni [54].[5]

The idea of using the inventions of living systems in the development of technical devices and innovations is more than a century old; there exists an extensive literature on the topic. A semiotic view may add the following: (1) it demonstrates the logical differences between the life-bound products and non-living assemblies; (2) by describing more profoundly how organisms search, find, and build, we can understand their cleverness (for instance, as it occurs, their mechanisms of development and evolution may differ considerably from the algorithms of optimization); and, (3) a semiotic analysis of bio-inspired technological development itself opens up additional cultural, economic and ecologically valuable aspects.

The relationship between bionics and semiotics was noticed by Sebeok [58: 555]: "Our knowledge of basic zoosemiotic processes may also be put to practical uses to supplement existing human information-handling devices, and to advance bionics, a term that designates a rapidly growing field which aims to develop nonliving systems on the analogy of biological information-storing, coding and sorting systems".

Hoffmeyer ([25, 26]; also Bruni [6]) took a further step by emphasizing the importance of developing biosemiotic technology. Hoffmeyer wrote: "we are now finally prepared to complete the project that was initiated in the late Middle Ages with the introduction of windmills and water mills—i.e., to set societal production free of the constraining bonds given by the peculiarities of organic life. The first time around, under the Industrial Revolution, only the dimension of *energy* was set loose from the constraints of organismic life. Therefore, what we are now facing—and to some extent have already engaged in—is the setting free (and harnessing) of the semiotic dimension from its bindings in organic life. This is what I have called the development of *biosemiotic technologies*" [26: 344].[6]

In recent years, the relationship between biosemiotics and biomimetics has been explicitly stated by several authors (e.g., [30, 63]). This does not concern only

[4]Assuming that the problem-solving by living systems also means some gaining of knowledge, we may shed light on these processes by mimicking life's findings. In a more general sense, mimicking can be a tool for investigation, since mimicking is a kind of modelling.

[5]Rossi and Pieroni [54] write: "In addition to the development of bioinspired artifacts for achieving better performance, another dimension of interest is epistemological. The epistemological approach attempts to test and verify biology-based hypothesis by conceiving and implementing specific bioinspired machines".

[6]Application of semiotics in biomimetics has more aspects. For instance, Camargo and Vega [8: 161] emphasise "the importance of the semiotic theory as an intermediary field in the transposition of natural phenomenon from their original biological field to computational field".

technology—semiotics of bio-inspiration demonstrates a growing popularity also in arts and architecture.[7]

1 Biomimetics and Its Relatives

The aim of biomimetics is construction, which links biomimetics to various biomechanical approaches. This assumes a knowledge and understanding of the mechanics of biological samples, and therefore needs to know the biological *logic* of construction.

Over the past five or six decades, there have been differing but overlapping approaches that have focused on living systems and construction discoveries particularly in the overlapping regions between the fields of biology and engineering.[8] These approaches are defined below in an approximate historical order.

Biomechanics is a term that has been used since late nineteenth century to denote the study of structure and function of biological systems by the means of mechanics.

Biontotechnics is a term and concept proposed in 1902 by Tornier [61] for the field of modifying organisms or for using them technologically [21: 23].

Biotechnics was the term used by Raoul Francé in the early twentieth century [17]. It has the same meaning as bionics (and biomimetics) which was coined many years later [50]. The term was also used at the same time by a biologist and urban planner Patrick Geddes [21: 32].

Synthetic biology is currently known as "the design and construction of biological parts, devices and systems, and the redesign of existing, natural biological systems for useful purposes" [28: 707]. This term was introduced over a hundred years ago in 1910 by Stéphane Leduc. As mentioned by Bensaude-Vincent [4], Leduc's program was both synthetic and biomimetic. According to Leduc [39: 147], there is no sharp division between inanimate nature and life.

Biotechnology was used as a term by Károly Ereky in 1919, defining it as the production of products from raw materials with the aid of living organisms, and for procedures for modifying living organisms according to human purposes.[9] According to the Convention on Biological Diversity (article 2), 'biotechnology' is "any technological application that uses biological systems, living organisms or derivatives thereof, to make or modify products or processes for specific use".

---

[7]On semiotic approach to mimesis and its relationship to modelling systems, see also [40] and [22].

[8]See also [18].

[9]See a history of biotechnology by R. Bud [7].

Bioengineering was coined by British scientist and broadcaster Heinz Wolff in 1954. According to one definition, the mission of bioengineering is "to create a fusion of engineering and the life sciences that promote scientific discovery and the invention of new technologies and therapies through research and education".[10]

A considerable turn and growth of interest towards biomimetics took place around the early 1960s. This was spawned in part by a series of Macy conferences in New York (1941–1960) in which cybernetics, including its biological and medical aspects, were discussed.

The term **biomimetics** was seemingly coined by Otto Schmitt in 1957 in his doctoral thesis [5: 1] and used in 1969 in the title of his paper [57]. It was defined as "the study of formation, structure, or function of biologically produced substances and materials (as enzymes or silk) and biological mechanisms and processes (as protein synthesis or photosynthesis) especially for the purpose of synthesizing similar products by artificial mechanisms which mimic natural ones" [54].

The term **bionics** came into use following the first symposium of bionics in 1960 in Dayton, Ohio, USA [60]. It is often used as a synonym of biomimetics. However, if seen as a sub-discipline of biomimetics as mentioned by Rossi and Pieroni [54: 1], ""bionics" is more related to robotics (having an emphasis on biologically based control and intelligence), ethology-based robotics (having an emphasis on constructing robot hardware based on animals), and biomimetic actuators and sensors."

Adaptronics is a study of biologically inspired materials. According to one definition, these are "material systems that have intelligence and life features integrated in the microstructure of the material system to reduce mass and energy and produce adaptive functionality" [31: 1]. The term has been in use since the 1960s.[11]

Artificial Life is a field of study as formulated by Christopher Langton in the 1980s. It is a modelling and simulating of life processes, together with realization of concrete systems [48].

Biognosis is the term suggested by Rustum Roy (in the early 1990s) and defined as *learning* from living systems (not simply mimicking their form), deriving knowledge from biology to materials science.[12]

Biomimicry is another synonym to biomimetics. As defined on the home page of the Biomimicry Institute[13] as "an approach to innovation that seeks sustainable solutions to human challenges by emulating nature's time-tested patterns and strategies. The goal is to create products, processes, and policies—new ways of living—that are well-adapted to life on earth over the long haul."

---

[10]https://bioengineering.stanford.edu (2015).

[11]The term 'adaptronics' appears, for instance, in *Bionics Symposium 1966: Short Paper Pre-Prints*, Dayton (Quashnock, Joseph M., dir., Air Force Systems Command, US Air Force, Wright-Patterson Air Force Base, Ohio), paper by Cary W. Armstrong 1966 (pp. 1–9), p. 1—which is earlier than assumed in Janocha [31: 5] that relates it to VDI Technology Centre that was established only in 1978.

[12]*MRS Bulletin*, March 1995, p. 48.

[13]https://biomimicry.org/what-is-biomimicry/ (2015).

Biorobotics is a subfield of robotics—making robots that emulate or simulate living biological organisms. Since 'robot' may mean quite wide variety of semi-autonomous devices, there is certainly an overlap with biomimetics.

Nanobiotechnology, or biologically informed nanotechnology converges with nanoscale biomimetics.

Bio-inspired design (also 'biologically inspired engineering') is again another name for biomimetics, however often with an emphasis of broader aspects of design [20, 62].

Biohybrid and the biomorphic systems approach describe a design or relationship that is even closer than mimicking by sharing the principles of living systems and their behavior. As Sarpeshkar [56: 252] mentions, "the field of biomorphic design suggests that we can mine the intellectual resources of nature to create devices useful to humans, just as we have mined her physical resources in the past. Such mining will require us to combine inspiration with perspiration and to understand how nature works with insight."

Biomaterials science or biomaterials engineering focuses on studies on function and design of biomaterials (for instance, bioceramics), and on how materials interact with living organisms.[14]

Physionics is a term combined from 'physiology' and 'electronics' (while 'bionics' from 'biology' and 'electronics'). As Giordan [19] argues, while bionics has paid attention mainly to morphological and anatomical aspects, physionics focuses on physiological and regulatory aspect of living systems.

Semionics is an important but a fairly seldom used term that is not about the materials or devices, but is about the building of signs. Thus this is, in some analogy with bionics, a theory of sign formation [13: 99], a study of the creation of new signs, or a technology of making signs (cf. [24]) on the basis of models used by living systems.

What we can learn from the diversity of these terms and approaches is that there has been a recurrent fascination with the construction of living systems that has resulted in a whole movements with specific aims, emphasis and ideology. The long-term parallel existence of these movements is interesting and is the overriding focus of biomimetics.

2 How Organisms Acquire and Preserve Their Materials, Structures, and Mechanisms

Key questions are: why are biological materials and processes as they are, and why are they different from the materials and processes outside of living realm? And, what kind of role does life play in the establishing these body-materials?

[14] On the contemporary material biomimetics, see [59].

The materials and processes that are technologically mimicked are of course not in themselves life. These are materials and structures that both enable life and are products of life. These are structures that allow life to proceed and are the scaffolding (channeling, limiting, restricting and constraining) of the life processes.

While biomimetics turns attention to the aspect of transferring the findings of living systems into a technology, the role of biosemiotics includes explanations of the innovative capacity of life, description of the ways life is doing its discoveries and polishing them, and also working out typologies and possible measures of the functional value of organic qualities. Biosemiotics is focused on the study of life itself.

When using various structures or mechanisms of living systems as paragons and as examples to follow and mimic, it is important to understand the ways and limits of perfection of biosystems themselves, yet before or without any cultural design. This is because the ways by which certain forms or structures have been achieved or constructed is usually life-dependent. Likewise, their chemical composition is also life-dependent.

A Russian biogeochemist Vladimir Vernadsky paid attention to the fact that living matter, i.e. the material the organisms are built of, is in several ways different from the matter that is produced without the assistance of life. For instance, there exists a wide diversity of organic substances and materials created by organisms. Analogically, when making a distinction between the biosphere and the human noosphere, Vernadsky pointed out chemical substances and materials that required only the human symbolic mind in order to be produced, such as free aluminium or molybdenium which did not occur in native form before our anthropocene.

Organisms' capacity to catalyse and construct is based on their acquired features. These features work as constraints that channel behaviour of organelles, cells and organisms. These learnt constraints persist via inheritance and can be interpreted as knowledge in a broad sense. One way biologists have spoken about the knowledge of organisms is via the concept of adaptation. In this sense, everything meaningful that organisms do or make or build is adaptational. However, one needs to be careful with this term because as it is used by almost all schools of thought in biology, the concepts behind this term may have very different definitions.

The neo-darwinian explanation of the origins of adaptations claims that adaptations are exclusively worked out by natural selection, that is by the differential replication of randomly occurring differentiation of forms. However, do the changes that occur as a result of natural selection create some knowing for the organism itself? Since the number of replicas is not a feature of a single replica (i.e., behaviour of a particular system does not depend on the number of copies of this system in the world), the differential reproduction of organisms cannot by itself change the information the organisms possess.

Nevertheless, the mechanism of natural selection (defined as the differential reproduction of genotypes) has been seen as a certain primitive optimization mechanism. Random mutations provide new genomes, replication reproduces and thus amplify some of the genomes (that are called "fit"); their persistence (stability) can be different. Also, due to the limitedness of resources there appears feedback that "compares" the efficiency of variants.

This mechanism, however, optimizes poorly. First, evolution is just so extraordinarily slow—the number of variants of genomes that has ever been created and thus tested throughout the whole history of life (roughly equal to the number of individuals lived during three billion years) is utterly small in comparison with the number of possible combinations of a gene. Second, even if an optimum is found, there is no reason to stay on it due to ever changing context or conditions (which are also changing under the influence of organisms themselves), including the ever changing other parts of the genome, and because many of the less efficient solutions can also survive.

Thus we can say that biological evolution does not have optimality as its general aim. An "arms race" is quite exceptional in most communities and species. The main drive in the world of life is the fulfilling of needs, while there is no general reason or tendency to go up to the limits. Furthermore, there are seemingly no simple general invariants or universals of biological perfection or progression.

It is obviously not enough to explain the innovation, or abduction (inference on the basis of hypothesis), via the amplification of random change, as the explanation via the mechanism of natural selection would assume and suggest. What can then be said about the mechanism which may lead to knowledge?

Nevertheless, in addition to natural selection, there exists another mechanism that works for finding solutions—direct choices made by organisms. These are the choices between options that are available simultaneously—of food, of directions, of construction, of signals, etc. This mechanism has been studied much less, mainly because it requires knowledge about the umwelt of the organism, about the distinctions organism can make, and especially about its intrinsic time. The nature of this mechanism is to solve the incongruences, to remove the logical conflicts, and thus to move towards logically more consistent processes. This is a fast mechanism. This is the field of epigenetic processes. Adaptive epigenetic processes can be called learning. However, there is no direct way from a solution found in behavior to the inheritance of that solution.[15]

Knowing can be defined as something that results from learning. Adaptations, in the biological sense, are such. This means that adaptations are carrying certain knowledge. Knowing, including its primary and primitive forms, expresses itself also as modelling. In this general sense, artefacts also carry meaning, i.e. knowing. All these are directly semiotic aspects of life.[16]

[15]On the mechanisms of how the behavioural decisions can be conveyed via epigenetic and ecological inheritance and further fixed through genetic drift, see, e.g., [35, 65].

[16]It should be emphasized that semiotics is dealing not only with sign processes or life processes, but also with what is built or constructed by life processes, like it is the case with artefacts in technology and structures of organism's body.

We can distinguish between three main stages in the process of learning:

(1) Incompatibility, or the situation of a problem. What can induce a functional change, is an incongruence of functional relations (i.e., of sign relations), a semiotic (logical) conflict or untranslatability, a confusion, a controversy, or a problem.[17]
(2) Innovation: if several options (potentialities) are available simultaneously, then choices have to be made and certain new links are established. This is problem-solving or decision-making. Here, the earlier experience in the form of constraints, or habits or scaffolding,[18] play a big role in channelling or directing the decision-making and creating innovation.
(3) Habituation: this is how the new connections or links may become stabilised, and sometimes, (partially) inherited.

In a general sense, functional relations are worked out or established to solve incompatibilities or problems. If a new functional relation is established, this can be called 'learning'. Learning, in this sense, is a broad concept; learning can occur as widely as semiosis.[19] Acquiring a functional relation can be of several types—recognition of something new, or association of something to something else, or imitation of some movement, making a new link, etc. As a functional relation, it is *for* something, i.e. for a certain use, aim, or purpose.

Establishing new links is, broadly speaking, the fundamental basis for building an artefact.[20] An organism, in this sense, is a network of functional links established by the life processes themselves. Thus, an organism's body is an artefact that has resulted from earlier learning. Everything in life, all its diversity of forms and processes, is a result of a continuous search with dialogues and negotiations during millions of years. Commonly, this is a permanent practice or usage of materials or structures (analogous to organs) for a very long time that involve many cycles of rebuilding and allow the structures and materials to stay stable.

Most of the functional changes take place in evolution in the form of exaptation, defined as the change in function of earlier existing structures. This is possible due to potential multifunctionality or polysemy of organic forms. Some changes have been inherited. For this to happen there must be scaffolds that delimit functioning considerably. This being the case, it is possible that despite of potentially wide range of usages, the structures are used mostly in the same way for so long time that the scaffolds could be fixed due to random drift in genome.[21]

---

[17]As we have argued, it is also where the organism's phenomenal world stems from and the phenomenal present appears. This is because conflict (as well as choice) presupposes options, but options assume simultaneity, which implies that if an organism makes choices, it should have a phenomenal present. See [38].

[18]See [37].

[19]Semiosis or sign process is also ordering—the process in which indeterminacy turns into a relation, and further into a habit or rule. Cf. [29: 132].

[20]Spelled also as 'artifact'.

[21]A more detailed description of this mechanism, see [35].

What makes an artefact a non-living semiotic structure are the bonds that make and hold this structure.[22] These bonds (which are built into artefacts and represent their embedded code[23]) are both dynamic and static. Dynamic bonds are generally of the *if* ... *then* logical type as a perception–action relation. Static bonds are glue-type bonds that are both replicable and non-replicable, one-time and unique bonds which hold together the pieces of built patterns. Artefacts cannot be constructed only on the basis of self-assembly, since their construction requires work.

Thus, life is intelligent because of its capacity to be illogical—incompatible and conflicting. This paradoxical formulation points to the situation on which knowledge-making stands: the situation of choice that is a situation of incompatibility of rules, a situation of confusion and a challenge to problem-solving. This is where abduction takes its origin and where invention becomes possible.

The materials, structures and mechanisms of living systems are smart because they store the results of the choices that have been made. Knowledge can be defined as traces of choices. Knowledge acquisition is based on short-term processes—decision making in situations of incompatibility and making choices if there are options. As a result, the systems are rebuilding and redesigning themselves. This can mean that innovations are quick, but also that fine-tuning takes a long time.

Natural selection (i.e., differential reproduction) has certainly played a role in filtering out the solutions and life's findings. However, the findings, innovations, inventions themselves require additional semiotic description of interpretations or of choices or decision-making. Thus the biosemiotic view in biomimetics would add the understanding as to why the solutions found in living systems can be useful and should be evaluated, and also how organisms create value by creating the systems that persistently work.

3 Perfection for Sustainability: Changing Without Trace (Some Ecosemiotic Aspects of Biomimetic Work)

With technology and with artefact-building we embed some knowledge into a matter. The embedded knowledge is a product of earlier choices and decisions. It works as scaffolding and constraints for the next situations of choice. With this, we influence the decision-making in the future. Some needs can be fulfilled easier, some with a bigger effort. With this, we either facilitate or render difficulty to subsequent ideologies.

---

[22]On the relationship between signs and tools, cf. [46].

[23]Code as (a rule, based on) a mediated correspondence is commonly a carrier of knowledge. However, it may not be the case for all codes. Namely, if a code is a result of (originates from) a purely random processes, i.e. if it is not a result of choices, then it may not by itself carry any knowledge. For instance, genetic code as the correspondence between nucleotide triplets and amino acids is a code, while seemingly without any knowledge to carry on (because it is not a product of choices).

Biomimetics is frequently assumed to be a means for sustainable or green technology. The ecological approach and the idea of sustainability have obviously enhanced the searches in this area. However, life-inspired technology is only a half of solution towards a sustainable life. Sustainable economy, in an ecological sense, requires the zero balance in all element cycles, i.e., the thermodynamically closed but not isolated system. In an ecosystem, it is largely assured by biodiversity. Thus, diversity is an additional aspect that should be taken into account in development of biomimetics. Such an approach can be called *systems biomimetics*.

Technology that mimics living nature may be assumed to be environment friendly (or green) by itself [45: 5; 20, 32, 33]. However, biomimicking may not be sufficient criteria for sustainability. As, for instance, John Barry [3] has suggested, "the explicit examination of what constitutes 'progress' is central to the task of rethinking green politics". One way to approach this is to make a clear distinction between progress and perfection, from which only the latter meets the criteria of sustainability.[24]

A large-scale application of biomimetics is an aspect of what is sometimes called posthumanities (e.g., [55: 94; 42]). This is where a fusion between technology and living body takes place, where artefactual constructions and ecological network turn into a swarm intelligence [49].

Heinz von Foerster [16] has described bio-logic as coalition-making. In the context of ecological technology, we can interpret it as making coalition with the ecological network of an ecosystem in which the particular culture is a part, facilitating the functions of sustainability. In this case, we already take the ecosystemic processes as a certain metamodel for the technological design, and this would mean that we could even define the ecological approach as biomimetic in large.

A paradoxical aspect of life is that it is not only a problem-solver, but also the single problem-creator. Repeated decisions lead to habits, or rules of behaviour and action that organize the processes of an ecosystem. Since these acquired rules are not universal, they can be mutually incompatible and thus lead to various kinds of unpredictable conflicts. These conflicts provide new problems for organisms to solve, and the more complex the living (eco)system becomes, the more numerous and diverse the problems become. An ecological approach would attempt to find the type of solutions that would avoid catastrophic or unsolvable situations. For instance, Allen Newell [43] has written: "since we want machines to help us solve problems, the more intelligent we are able to make it, the more unobtrusive it

[24]This discussion can make a reference to the final (eleventh) thesis in Karl Marx' "Theses on Feuerbach" (1845), which sounds: "Philosophers have hitherto only *interpreted* the world in various ways; the point is to *change* it" (translation by Cyril Smith 2002). While the emphasis of Marx was just on practice, the thesis has been used as a call for an overwhelming and unlimited technical progress, a call to remake our environment on the basis of technical and industrial innovations. This principle, however, is not an implication from a model of sustainable ecosystem. Since the understanding of ecology of the biosphere, however, the major challenge of science became to be "how *not to change* the world".

should be in providing this help. Contemplating the more extreme forms of this vision, there is little to describe about the machine except that it gives a great deal of help quickly and with very little pain." Thus the biomimetic technology is both helping to solve ecological problems as well as create unpredictable new ones. An *ecosemiotic* view could be usable in order to understand these aspects of our activity [34, 41].

Living systems, except humans, do not define the aim of their construction or behaviour. They just solve or remove the inconsistencies by using some earlier experience in the form of habits or memories if they are available. The buildings this living process (semiosis) results are often amazing.

In a balanced ecosystem, everything what is produced is consumable to somebody in the same ecosystem—and it will be consumed. This illustrates an important and simple guideline for sustainability: the reversibility[25] or reusability of materials from the same ecosystem, together with low energy-use. In more semiotic terms, the features like non-universality, locality, communicative restrictedness, and context-dependence appear to be relevant in the building of scaffoldings for the sake of smarter choices.[26] Inspired by and in tune with one's own local ecosystem this may have an effect on improving health.[27]

Acknowledgements I thank Meir Israelowitz for encouraging me to write this essay, Timo Maran for helpful comments, and the project PRG314 that allowed completing this work.

References

1. Abir-Am, P.G.: The biotheoretical gathering, trans-disciplinary authority and the incipient legitimation of molecular biology in the 1930s: new perspective on the historical sociology of science. Hist. Sci. **25**, 1–70 (1987)
2. Bar-Cohen, Y. (ed.): Biomimetics: Nature-Based Innovation. CRC Press, Taylor & Francis Group, Boca Raton (2012)
3. Barry, J.: Rethinking Green Politics: Nature, Virtue and Progress. Sage, London (1999)
4. Bensaude-Vincent, B.: Biomimetic chemistry and synthetic biology: a two-way traffic across the borders. HYLE: Int. J. Philos. Chem. **15**(1), 31–46 (2009)
5. Bhushan, B.: Biomimetics: Bioinspired Hierarchical-Structured Surfaces for Green Science and Technology. Springer, Heidelberg (2012)
6. Bruni, L.E.: A sign-theoretic approach to biotechnology. Thesis, University of Copenhagen, Copenhagen (2003)
7. Bud, R.: The Uses of Life: A History of Biotechnology. Cambridge University Press, Cambridge (1993)
8. Camargo, C.E.P., Vega, I.S.: Semiotic transposition method for biomimetic software modeling: an overview. In: Arabnia, H.R., Deligiannidis, L., You, J. (eds.) Proceedings of the 2014 International Conference on Modeling, Simulation and Visualization Methods, pp. 155–161. CSREA Press, Las Vegas (2014)

[25]On certain irreversibility of growth of knowledge, however, see [47].

[26]On some other semiotic qualities of biomimetic product design, see, e.g., [9, 15].

[27]Cf. [27].

9. Coelho, D.A., Versos, C.A.M., Silva, A.S.C.: Product personality assignment as a mediating technique in biologically and culturally inspired design. In: Ji, Y.G. (ed.) Advances in Affective and Pleasurable Design, pp. 361–370. CRC Press, Boca Raton (2012)
10. Cutsuridis, V., Hussain, A., Taylor, John G. (eds.): Perception-Action Cycle: Models, Architectures, and Hardware. Springer, New York (2011)
11. Emmeche, C.: The chicken and the Orphean egg: on the function of meaning and the meaning of function. Sign Syst. Stud. **30**(1), 15–32 (2002)
12. Emmeche, C., Kull, K. (eds.): Towards a Semiotic Biology: Life is the Action of Signs. Imperial College Press, London (2011)
13. Epstein, M.: The Transformative Humanities: A Manifesto (Klyukanov, I., trans.) Bloomsbury, New York (2012)
14. Favareau, D. (ed.): Essential Readings in Biosemiotics: Anthology and Commentary. Springer, Dordrecht (2010)
15. Figueiredo, J.F.D., Coelho, D.A.: Semiotic analysis in perspective: a frame of reference to inform industrial design practice. Des. Principles and Prac. Int. J. **4**(1), 333–346 (2010)
16. Foerster, H.: Bio-logic. In: Bernard, E.E., Kare, M.R. (eds.) Biological Prototypes and Synthetic System, vol. 1. (Proceedings of the Second Annual Bionics Symposium, Cornell University, 30 Aug–1 Sept, 1961), pp. 1–12. Plenum Press, New York (1962)
17. Francé, R.H.: Plants as Inventors. Simkins and Marshall, London (1926); Die Pflanze als Erfinder. Kosmos, Stuttgart, Gesellschaft der Naturfreunde (1920)
18. Gebeshuber, I.C., Drack, M.: An attempt to reveal synergies between biology and mechanical engineering. Proc. Inst. Mech. Eng. Part C: J. Mech. Eng. Sci. **222**(7), 1281–1287 (2008)
19. Giordan, A.: Voici venue l'ère de la physionique. La Recherche **27**, 80–86 (1996)
20. Goel, A.K., Bras, B., Helms, M., Rugaber, S., Tovey, C., Vattam, S., Weissburg, M., Wiltgen, B., Yen, J.: Design patterns and cross-domain analogies in biologically inspired sustainable design. In: Proceedings of the AAAI Spring Symposium on AI and Sustainable Design, pp 45–52. Stanford University, Palo Alto (2011)
21. Goujon, P.: From Biotechnology to Genomes: The Meaning of the Double Helix. World Scientific Publishing, Singapore (2001)
22. Grigorjeva, J.: Lotman on mimesis. Sign Syst. Stud. **31**(1), 217–237 (2003)
23. Gruber, P., Bruckner, D., Hellmich, C., Schmiedmayer, H.B., Stachelberger, H., Gebeshuber, I.C. (eds.): Biomimetics: Materials, Structures and Processes: Examples, Ideas and Case Studies. Springer, Berlin (2011)
24. Gudwin, R.R.: Semionics: a proposal for the semiotic modeling of organizations. In: Liu, K. (ed.) Virtual, Distributed and Flexible Organisations: Studies in Organisational Semiotics, pp. 15–33. Kluwer Academic Publishers, Dordrecht (2004)
25. Hoffmeyer, J.: S/E > 1: A semiotic understanding of bioengineering. Sign Syst. Stud. **29**(1), 277–290 (2001)
26. Hoffmeyer, J.: Biosemiotics: An Examination into the Signs of Life and the Life of Signs. University of Scranton Press, Scranton (2008)
27. Hoffmeyer, J.: A biosemiotic approach to health. In: Cowley, S.J., Major, J.C., Steffensen, S. V., Dinis, A. (eds.) Signifying Bodies: Biosemiosis, Interaction and Health, pp. 21–41. The Faculty of Philosophy of Braga Portuguese Catholic University, Braga (2010)
28. Holm, S.: Organism and artifact: proper functions in Paley organisms. Stud. Hist. Philos. Biol. Biomed. Sci. **44**(4, part B), pp 706–713 (2013)
29. Innis, R.E.: Dimensions of an aesthetic encounter. In: Gertz, S.K., Valsiner, J., Breaux, J. P. (eds.) Semiotic Rotations: Modes of Meanings in Cultural Worlds, pp. 113–134. Information Age Publishing, Charlotte (2007)
30. Ireland, T.: A cell-inspired model of configuration. In: Combs, L., Perry, C. (eds.) Computational Ecologies: Design in the Anthropocene. Proceedings of the 35th Annual Conference of the Association for Computer Aided Design in Architecture (ACADIA), pp. 137–147. University of Cincinnati, Cincinnati (2015)
31. Janocha, H. (ed.): Adaptronics and Smart Structures: Basics, Materials, Design, and Applications. Springer, Berlin (1999)

32. Jorna, R. (ed.): Sustainable Innovation: The Organisational, Human and Knowledge Dimension. Greenleaf Publishing, Sheffield (2006)
33. Kennedy, E.B., Fecheyr-Lippens, D.C., Hsiung, B.K., Niewiarowski, P.H., Kolodziej, M.E.: Biomimicry: a path to sustainable innovation. Des. Issues **31**(3), 66–73 (2015)
34. Kull, K.: Semiotic ecology: different natures in the semiosphere. Sign Syst. Stud. **26**, 344–371 (1998)
35. Kull, K.: Adaptive evolution without natural selection. Biol. J. Lin. Soc. **112**(2), 287–294 (2014a)
36. Kull, K.: Zoosemiotics is the study of animal forms of knowing. Semiotica **198**, 47–60 (2014b)
37. Kull, K.: Evolution, choice, and scaffolding: semiosis is changing its own building. Biosemiotics **8**(2), 223–234 (2015a)
38. Kull, K.: Semiosis stems from logical incompatibility in organic nature: why biophysics does not see meaning, while biosemiotics does. Prog. Biophy. Mol. Biol. **19**(2) (2015b)
39. Leduc, S.: The Mechanism of Life. William Heinemann, London (1914) [1912]
40. Maran, T.: Mimesis as a phenomenon of semiotic communication. Sign Syst. Stud. **31**(1), 191–215 (2003)
41. Maran, T., Kull, K.: Ecosemiotics: main principles and current developments. Geografiska Annaler: Ser. B, Hum. Geogr. **96**(1), 41–50 (2014)
42. Merrell, F.: Sensing Corporeally: Toward a Posthuman Understanding. University of Toronto Press, Toronto (2003)
43. Newell, A.: On programming a highly parallel machine to be an intelligent technician. In: Proceedings of the Western Joint Computer Conference, May, vol. 17, pp. 267–282 (1960)
44. Niemann, H.J.: Karl Popper and the Two New Secrets of Life: Including Karl Popper's Medawar Lecture 1986 and Three Related Texts. Mohr Siebeck, Tübingen (2014)
45. Nosonovsky, M., Bhushan, B. (eds.): Green Tribology: Biomimetics, Energy Conservation and Sustainability. Springer, Heidelberg (2012)
46. Nöth, W.: On the instrumentality and semiotic agency of signs, tools, and intelligent machines. Cybern. Hum. Knowing **16**(3/4), 11–36 (2009)
47. Nöth, W.: The growth of signs. Sign Syst. Stud. **42**(2/3), 172–192 (2014)
48. Packard, N.H., Bedau, M.A.: Artificial life. In: Nadel, L. (ed.) Encyclopedia of Cognitive Science, vol. 1, pp. 209–215. Nature Publishing Group, London (2003)
49. Parikka, J.: Insect Media: An Archaeology of Animals and Technology (Posthumanities 11). University of Minnesota Press, Minneapolis (2010)
50. Pichler, F.: The contribution of Raoul H. Francé: biocentric modeling. In: Weibel, Peter (ed.) Beyond Art: A Third Culture: A Comparative Study in Cultures, Art and Science in 20th Century Austria and Hungary, pp. 371–373. Springer, Wien (2005)
51. Pohl, G., Nachtigall, W.: Biomimetics for Architecture and Design: Nature—Analogies—Technology. Springer, Cham (2015)
52. Popper, K.: All Life is Problem Solving (Camiller, P., trans.) Routledge, London (1999) [1994]
53. Popper, K.: A world without natural selection but with problem solving. In: Niemann, H. J. (ed.) Karl Popper and the Two New Secrets of Life: Including Karl Popper's Medawar Lecture 1986 and Three Related Texts, pp. 132–133. Mohr Siebeck, Tübingen (2014)
54. Rossi, D., Pieroni, M.: Grand challenges in bionics. Front. Bioeng. Biotech. **1**(3), 1–3 (2013)
55. Santaella, L.: Posthuman culture. In: Gertz, S.K., Valsiner, J., Breaux, J.P. (eds.) Semiotic Rotations: Modes of Meanings in Cultural Worlds, pp. 93–112. Information Age Publishing, Charlotte (2007)
56. Sarpeshkar, R.: Neuromorphic and biomorphic engineering systems. McGraw-Hill Yearbook of Science and Technology 2009, pp. 250–252. McGraw-Hill, New York (2009)
57. Schmitt, O.H.: Some interesting and useful biomimetic transforms. In: Third International Biophysics Congress of the International Union for Pure and Applied Biophysics (Boston, Mass., 29 Aug–3 Sept, 1969): Abstracts, p. 297. MIT, Cambridge (1969)

58. Sebeok, T.A.: On chemical signs. In: Kristeva, J., Rey-Debove, J., Umiker, D. (eds.) Essays in Semiotics. Essais de sémiotique. (Approaches to Semiotics, 4.), pp. 549–559. Mouton, The Hague (1971)
59. Shimomura, M.: The new trends in next generation biomimetics material technology: learning from biodiversity. Sci. Technol. Trends—Q. Rev. **37**, 53–75 (2010)
60. Steele, J.E.: How do we get there. In: Gray, C.H. (ed.) Bionics Symposium: Living Prototypes—The Key to New Technology, pp. 55–60. Routledge, New York (1960)
61. Tornier, G.: Ueberzählige Bildungen und die Bedeutung der Pathologie für die Biontotechnik (mit Demonstrationen). In: Matschie, P. (ed.) Verhandlungen des V. Internationalen Zoologen-Congresses zu Berlin, 12–16 Aug 1901, pp. 467–500. Gustav Fischer, Jena (1902)
62. Vincent, J.F.V., Bogatyreva, O.A., Bogatyrev, N.R., Bowyer, A., Pahl, A.K.: Biomimetics: its practice and theory. J. R. Soc. Interface **3**(9), 471–482 (2006)
63. Weaver, T.: Biomimetics: emulation and propagation in post-traditional ecologies. In: Hughes, R., Sundén, J. (eds.) Second Nature: Origins and Originality in Art, Science, and New Media, pp. 205–225. Axl Books, Stockholm (2011)
64. Webb, B.: Can robots make good models of biological behaviour? Behav. Brain Sci. **24**(6), 1033–1050 (2001)
65. West-Eberhard, M.J.: Developmental Plasticity and Evolution. Oxford University Press, Oxford (2003)

Medical Biotechnology and Biomimetics: Prospects and Challenges in Sub-Saharan Africa

Obaro S. Michael

Abstract Advancements at the interphase of engineering and biology have produced disciplines that include Biotechnology and Biomimetics. These new disciplines proffer exciting solutions to real life problems. New technologies are available for the purification of water and treatment of sewage; challenges that are common in resource-limited countries. In the medical sciences, tissue engineering methods are rapidly and positively transforming the field of regenerative and restorative Medicine. The importance of Biotechnology and Biomimetics in the management of infectious diseases and in drug development (Pharmacology and Toxicology) is rising exponentially. However, for sub-Saharan Africa (SSA), these advances are yet to make meaningful and significant impact. Biotechnology and Biomimetics are still at very rudimentary stages of development with many challenges militating against progress. The prospects and challenges for SSA are discussed here.

Keywords Infectious diseases · Tissue engineering · Sub-Saharan africa

1 Introduction

Biomimetics or biomimicry is the imitation of the models, systems, and elements of nature for the purpose of solving complex human problems. Drawing largely from physical sciences, the rapidly growing discipline has made significant contributions to large and small scale engineering. Nature has always inspired innovations and creativity [1, 2]. Over billions of years, nature has evolved and overcome challenges that humans are faced with for reasons that include expansion of human populations and migrations to uninhabited lands. From aerodynamics, to the building of submarines, man has copied nature. These awesome engineering feats have inspired

O. S. Michael (✉)
Department of Pharmacology and Therapeutics, Faculty of Basic Medical Sciences, College of Medicine, University of Ibadan, Ibadan, Nigeria
e-mail: micobaro@yahoo.com

M. Israelowitz et al. (eds.), *Biomimetics and Bionic Applications with Clinical Applications*, Series in BioEngineering,
https://doi.org/10.1007/978-3-319-53214-1_2

more and more discoveries into the inner workings of physical and biological phenomena. More recently, Biomimetics has found more significance in the medical field which holds real and potential benefits for all mankind.

Africa is the second largest continent of the world with a population of 1.136 billion as at mid-2014, Fig. 1 [3]. Sub-Saharan Africa (SSA) is home to a large proportion of the countries in the continent, almost all of them resource-limited. The burden of infectious diseases like malaria, Tuberculosis, HIV, respiratory infections, malnutrition, parasitic diseases, and rapidly rising rates of chronic diseases (Diabetes, cancer) is disproportionately huge. This is a region that stands to benefit greatly from the new discipline of medical biomimetics; the medical applications of biomimetics. This chapter discusses briefly the potential applications of the discipline in sub-Saharan Africa where a lot of peculiarities hamper research and development, and also discusses some of the challenges to technological advancements. Biomimetics has a role to play in our sub-region, but it is wise to examine some approaches that may help overcome the present challenges militating against changes that African traditional systems often perceive as threats to traditional cultural lifestyles.

The potentials of biomimetics research and development in Africa will be in the areas of technologies that will provide cheap methods of producing clean water, disposal of biological waste including sewage, and diagnosis and monitoring of therapeutic approaches to infectious diseases. In this write up, the discussion covers approaches to water and sewage treatment, diagnosis of malaria, HIV, and tuberculosis, special technologies for estimation of drug therapy and drug delivery,

Fig. 1 Africa

engineering bone and skin tissues for management of trauma and extensive burns. Challenges to this approach include poor power supply, lack of infrastructure, socio-cultural resistance to technological advancements, educational systems that do not incorporate the development and deployment of local technology, and mal-adaptation of foreign technology. The current approach of importing foreign technology and planting them on existing traditional systems, without adaptation, has not been sustainable. Thus, there are many institutions with expensive equipment that are lying abandoned because the infrastructure cannot sustain them. Some solutions are proffered so that the development of biomimetics will factor in peculiarities of sub-Saharan Africa, by developing approaches that are sustainable and adapted to the sub-region.

2 Water and Sewage in Sub-Saharan Africa

Water is an essential requirement of life and it drives biological processes in all lifeforms on Earth. Getting clean water for domestic and industrial use has remained a luxury in many communities of sub-Saharan Africa (SSA). Diseases borne by water still rank amongst the major risk factors for morbidity and mortality in sub-Saharan Africa [4]. While it is energy consuming, and thus expensive to purify water traditionally (by boiling or distillation systems), recent advances in biomimetics may solve this challenge. Biomimetics research into water and ionic pore systems have the potential of effective separation of molecules into pure forms. Initial studies have shown the possibility of detecting low weight micro-organisms, heavy metals, and very low concentration of other bio-organic molecules, which are often the contaminants in many sources of water [5, 6]. Using what can be called an electronic nose; contaminants in water may be readily detected and planned for purification processes. The mimicking of water purification systems in plants holds the potential of developing low energy water purifications methods. For a long time, bacteria have been used to decontaminate sewage and recycle water [7, 8]. However, African traditional systems shy away from such natural systems considered 'unclean.'

3 Biotechnology and Infectious Diseases

Infectious disease significantly affects the economic and socio-cultural activities of nations. There is room for improvements in the detection and diagnosis of the causative agents of Malaria, Tuberculosis, and the Human immunodeficiency virus (HIV). With improvements in biological sensors, this application of biotechnology is becoming more and more feasible. Micro-disc electrode arrays are implantable devices that can monitor biochemical pathways in the body [9]. The malaria pigment hemozoin, and its metabolism, has increasingly become a potential target for development of biotechnological methods of detecting the malaria parasite and potential drug targets non-invasively [10, 11]. With improvements in biomimetic recognition of protein, carbohydrates, and nucleic acids, diagnosis of bacteria (e.g. mycobacterium tuberculosis), and viruses (e.g. HIV), will become simpler and automated [12]. Cell targeting and gene transfers are technologies that can transform the field of infectious diseases [13]. Biosensors may someday be the technological basis of new real time diagnostic methods [14]. These are huge leaps in the medical field and are a few technologies that can change the terrain of management of infectious diseases; the major plagues of the African continent.

Biotechnology offers the possibility of rapid diagnosis of diseases. Chronic illnesses that are prevalent in developed countries are currently hot areas of research. Rapid improvements in infra-red camera technology has already found a place in the characterization on non-viable tissue in major limb neuropathy and ischemia due to diabetes [15–20]. While medical applications of biotechnology in chronic diseases prevalent in developed countries is a much researched topic, the importance of biotechnology still remains largely a theoretical aspect of tropical medicine. Devices that can diagnose infectious diseases earlier, and monitor the pathogens non-invasively are urgently needed in Africa.

The polymerase chain reaction has become a most important method of diagnosis of infectious and non-infectious deceases [21–24]. Green florescent protein is increasingly being used to tag important pathogenic pathways [25]. The culture of cells and micro-organisms will be further developed using bioreactor technologies that can maintain controlled internal conditions for optimal growth and development of cells [26, 27]. These methods have led to unprecedented insights into human pathogenic processes. The capacity for high throughput research needs to be developed, while challenges to technology in the continent must be surmounted.

4 Tissue Engineering

Advances in tissue engineering are providing new therapies for damaged human tissues. Bone and skin are commonly affected in traumatic accidents. In Africa road traffic accidents, major burns, and limb amputations due to complications of diabetes and other diseases are common. Developments in tissue engineering are potential solutions to these physically and emotionally distressing conditions. Biotechnology and biomimetics have made very promising devices and gadgets that can replace loss of human tissues. Artificial heart valves, dialysis machines, different forms of prostheses are life-saving, available readily in developed nations; with some difficulties, these devices are also obtainable in the African continent.

The use of biomimetic systems represents a major area of research in implant surgery. Developing artificially grown bone, teeth, and skin using biomimetic approaches will offer opportunities for restorative medicine and surgery [28, 29]. There are attempts at engineering whole organ systems including the heart and lungs. Nervous tissue is being coerced to regenerate using methods learnt from nature. The diseases of ageing, a good example being Alzheimer's disease, remain incurable, but developments in biotechnology are positive and encouraging. Artificial limbs for amputees, hearts for end stage heart disease, replacement of failed kidneys, and new lungs for those with end stage chronic obstructive lung diseases. All these point to a future that is full of promise; by mimicking nature we may advance to a point where out technology will reverse pathologies that are largely considered irreversible. However, there is very little or no visible research activity along these lines in sub-Saharan Africa to date. Efforts are being made, but much resistance and adverse conditions are mitigating progress.

4.1 Applications of Biotechnology to Pharmacology and Toxicology

Recent advances in biotechnology and bioinformatics have introduced very powerful tools for drug development. Synthetic mimics of endogenous ligands are being explored. One promising area is the management of diabetes where the search for insulin mimetics is being vigorously researched [30]. The scale up in the production of synthetic of artemisinins will reduce the cost of treatment of malaria in sub-Saharan Africa [31]. Antibiotics, antihelminthics, hormones, cancer therapy, and other bioactive molecules are some positive yields of advances in biotechnology and biomimetics [32–36]. Yet there is still much more in plants with potential benefits in human medicine.

Africa has a rich herbal medicines industry but this has remained largely unrefined and as crude extracts. Herbal and traditional medicines in Africa have not benefited from modern tools of biotechnology. Recent efforts at sequencing of genomes of plants, animals, and humans will best be utilized in the continent that harbours a high frequency of genetic polymorphisms. While herbs are generally considered safe, interactions with orthodox medicines are just being researched. In addition, toxicities of herbal medicines are also being discovered [37]. The new tools that are being developed in biotechnology are needed to refine traditional and herbal medicines. Genetic transformation is a powerful tool for enhancing the productivity of secondary plant metabolites and bioreactors are key steps towards commercial production [38]. In China modern tools of genetic engineering are provoding means of breeding of new medicinal plant varieties with high and stable yield, good quality, as well as stress-resistance [39]. These technologies are required in Africa where there is a rich culture of traditional and herbal medicine practice. However, there is overwhelming evidence that there are real challenges that have hampered progress, and continue to slow down attempts at transforming traditional medical practices in sub-Saharan Africa.

5 Challenges in Sub-Saharan Africa

Nature has perfected natural phenomena over billions of years. As exciting as this observation is, it is even more exciting to realise that what nature has built over the millennia, human technology may be able to improve upon and automate. Technological advances in recent times have transformed biology and medicine. Diseases that were previously considered incurable are being detected earlier and being cured. This marriage of engineering and biology is on a rapid rise in developed countries. Unfortunately sub-Saharan Africa (SSA) is lagging far behind in this activity; an activity that holds the most benefit to SSA, perhaps more than to any other region of the world. There are many reasons why this is so, but three will be mentioned: inherent inertia and resistance to change of African traditional systems, general absence of attitude of responsibility and respect for human intellectual effort, and poor infrastructural development.

Across the continent of Africa, in countries of sub-Saharan Africa, African traditional systems co-exist with orthodox, largely, western settings. The majority of the people are rural and have traditional religious and socio-cultural systems that pre-date modern history. With the advent of western education, urban dwellings based on western culture have continued to expand; however, the traditional beliefs are strongly entrenched even in the minds of the educated. One attribute of traditional African systems is resistance to change and often encountered ritualistic practices. The young are not raised to question what has existed or has been inherited. This is a major factor that has limited technological advancements in the sub-region. The use of objective evidence to drive social, cultural, and infrastructural changes in the society is foreign to such settings. Nevertheless, there is hope that, with rising numbers of educated Africans, these practices will not remain for too long. There still remains the need to build into the curriculum of primary schools the need to develop critical thinking, frequent challenge of existing bodies of knowledge, and questioning of effectiveness of long standing traditional systems. There should be no hesitation whatsoever in discarding old systems that are inimical to development.

Another challenge that has mitigated against technological advancements in many African settings is mis-interpretations and mis-applications of religion. The average African is very religious and in religion he or she finds self-dignity. There is no harm in being religious, in fact religions may offer some form of psychological stability, provide moral codes for those who practice different faiths, and may even have some health benefits [40]. These moral codes have in a large extent offered different degrees of social order and security. However, and rather difficult to understand, social responsibility has in recent times been increasingly negated by certain beliefs. The belief that progress comes only from the Divine leads to a form of neglect of social responsibility. When the people believe that God is the one who will rescue them from disease, unfavorable environments, poverty, and famine, not much else can be done. Long term planning is absent, and human intellectual effort is not praised, encouraged, or stimulated. There comes a general lack of curiosity and drive for discovery. Rather than religion inspiring positive actions aimed at producing beneficial changes, it becomes a passive resignation to fate which greatly hampers commitment to efforts aimed at technological and social development.

Finally, drawing directly from the above, long lasting neglect of active planning and good government that improve societal structures have led to huge infrastructural inadequacies. There are no structures or systems that can support the technologies discussed in this chapter. Electricity is absent or erratic in many sub-Saharan African countries [41]. Thus, when machines needed for state-of-the-art research are brought into sub-Saharan Africa, there is a high chance that they will rapidly become redundant and abandoned. Human capacity is also very inadequate; all these continue to fuel a vicious cycle that has become very difficult to break.

Thus, Africa lags behind in technological advances, despite the huge needs for the products. Disease is rampant and poverty continues to increase in the sub-region. Nigeria, the most populous sub-Saharan African nation has a disproportionately large burden of infectious and non-infectious diseases. The country has the highest burden of malaria, and a high burden of HIV [42–44]. However, there is hope. The continent has continued to improve socio-economically, and many sub-Saharan nations have grown educationally, medically, and economically. The recent elections in Nigeria (the 2015 elections) have been generally praised as a victory for democracy. Truly democratic nations are positioned to supporting biotechnological advancements. There is also increasing support from western nations towards empowering researchers of the region for high impact medical research. There are conferences that have been held on bringing biotechnological advances into Africa, and countries are putting in place policies that will encourage and guide developments in biotechnology [45–48]. These are very encouraging developments. Progress may be slow; but with sustained efforts, eventually sub-Saharan Africa will break into biotechnology, biomimetics, and be poised for relief from many burdens of disease.

Acknowledgements The authour acknowledges the valuable contributions of Dr. (Mrs) Afieharo Michael and Dr. Taiwo Lawal of the Department of Surgery, Faculty of Clinical Sciences, University of Ibadan, Nigeria. The author declares no conflict of interest.

References

1. Bar-Cohen, Y.: Biomimetics–using nature to inspire human innovation. Bioinspir. Biomim. **1** (1), P1–P12 (2006)
2. Bhushan, B.: Biomimetics: lessons from nature–an overview. Philos. Trans. A Math. Phys. Eng. Sci. **2009**(367), 1445–1486 (1893)
3. Bureau, P.R.: World population data sheet (2014)
4. Lim, S.S., Vos, T., Flaxman, A.D., Danaei, G., Shibuya, K., Adair-Rohani, H., Amann, M., Anderson, H.R., Andrews, K.G., Aryee, M., et al.: A comparative risk assessment of burden of disease and injury attributable to 67 risk factors and risk factor clusters in 21 regions,

1990-2010: a systematic analysis for the global burden of disease study 2010. Lancet **380** (9859), 2224–2260 (2012)
5. Canhoto, O.F., Magan, N.: Potential for detection of microorganisms and heavy metals in potable water using electronic nose technology. Biosens. Bioelectron. **18**(5–6), 751–754 (2003)
6. Wang, S., Xu, Z., Fang, G., Zhang, Y., Liu, B., Zhu, H.: Development of a biomimetic enzyme-linked immunosorbent assay method for the determination of estrone in environmental water using novel molecularly imprinted films of controlled thickness as artificial antibodies. J. Agric. Food Chem. **57**(11), 4528–4534 (2009)
7. Thomas, J.M., Ward, C.H.: Biotechnology for in situ restoration of ground water contaminated by the petroleum industry. Schriftenr. Ver. Wasser. Boden. Lufthyg. **80**, 345–365 (1989)
8. Janssen, D.B., Witholt, B.: Developments in biotechnology of relevance to drinking water preparation. Sci. Tot. Environ. **47**, 121–135 (1985)
9. Justin, G., Finley, S., Abdur Rahman, A.R., Guiseppi-Elie, A.: Biomimetic hydrogels for biosensor implant biocompatibility: electrochemical characterization using micro-disc electrode arrays (MDEAs). Biomed. Microdevices **11**(1), 103–115 (2009)
10. Egan, T.J.: Haemozoin formation. Mol. Biochem. Parasitol. **157**(2), 127–136 (2008)
11. Egan, T.J.: Recent advances in understanding the mechanism of hemozoin (malaria pigment) formation. J. Inorg. Biochem. **102**(5–6), 1288–1299 (2008)
12. Walker, D.B., Joshi, G., Davis, A.P.: Progress in biomimetic carbohydrate recognition. Cell. Mol. Life Sci. **66**(19), 3177–3191 (2009)
13. Wang, Y., Mangipudi, S.S., Canine, B.F., Hatefi, A.: A designer biomimetic vector with a chimeric architecture for targeted gene transfer. J. Control Release **137**(1), 46–53 (2009)
14. Jelinek, R., Silbert, L.: Biomimetic approaches for studying membrane processes. Mol. Biosyst. **5**(8), 811–818 (2009)
15. Ring, F.: Thermal imaging today and its relevance to diabetes. J. Diabetes Sci. Technol. **4**(4), 857–862 (2010)
16. Parker, M.D., Taberner, A.J., Nielsen, P.M.: A thermal stereoscope for surface reconstruction of the diabetic foot. Conf. Proc. IEEE Eng. Med. Biol. Soc. **2011**, 306–309 (2011)
17. Pafili, K., Papanas, N.: Thermography in the follow up of the diabetic foot: best to weigh the enemy more mighty than he seems. Expert Rev. Med. Devices 2014, 1–3 (2014)
18. Anburajan, M., Sivanandam, S., Bidyarasmi, S., Venkatraman, B., Menaka, M., Raj, B.: Changes of skin temperature of parts of the body and serum asymmetric dimethylarginine (ADMA) in type-2 diabetes mellitus Indian patients. Conf. Proc. IEEE Eng. Med. Biol. Soc. **2011**, 6254–6259 (2011)
19. Balbinot, L.F., Canani, L.H., Robinson, C.C., Achaval, M., Zaro, M.A.: Plantar thermography is useful in the early diagnosis of diabetic neuropathy. Clinics **67**(12), 1419–1425 (2012)
20. Balbinot, L.F., Robinson, C.C., Achaval, M., Zaro, M.A., Brioschi, M.L.: Repeatability of infrared plantar thermography in diabetes patients: a pilot study. J. Diabetes Sci. Technol. **7** (5), 1130–1137 (2013)
21. Gardner, A.F., Kelman, Z.: DNA polymerases in biotechnology. Front. Microbiol. **5**, 659 (2014)
22. Hamilton, S.C., Farchaus, J.W., Davis, M.C.: DNA polymerases as engines for biotechnology. Biotechniques **31**(2), 370–376, 378–380, 382–373 (2001)
23. Kranaster, R., Marx, A.: Engineered DNA polymerases in biotechnology. ChemBioChem **11** (15), 2077–2084 (2010)
24. Lovatt, A.: Applications of quantitative PCR in the biosafety and genetic stability assessment of biotechnology products. J. Biotechnol. **82**(3), 279–300 (2002)
25. Misteli, T., Spector, D.L.: Applications of the green fluorescent protein in cell biology and biotechnology. Nat. Biotechnol. **15**(10), 961–964 (1997)
26. Lavrentieva, A., Hatlapatka, T., Neumann, A., Weyand, B., Kasper, C.: Potential for osteogenic and chondrogenic differentiation of MSC. Adv. Biochem. Eng. Biotechnol. **129**, 73–88 (2013)

27. Weyand, B., Israelowitz, M., von Schroeder, H.P., Vogt, P.M.: Fluid dynamics in bioreactor design: considerations for the theoretical and practical approach. Adv. Biochem. Eng. Biotechnol. **112**, 251–268 (2009)
28. Bellamy, K.E., Waters, M.G.: Designing a prosthesis to simulate the elastic properties of skin. Biomed. Mater. Eng. **15**(1–2), 21–27 (2005)
29. Ashammakhi, N., Ndreu, A., Nikkola, L., Wimpenny, I., Yang, Y.: Advancing tissue engineering by using electrospun nanofibers. Regen. Med. **3**(4), 547–574 (2008)
30. Adachi, Y., Yoshida, J., Kodera, Y., Kato, A., Yoshikawa, Y., Kojima, Y., Sakurai, H.: A new insulin-mimetic bis(allixinato)zinc(II) complex: structure-activity relationship of zinc(II) complexes. J. Biol. Inorg. Chem. **9**(7), 885–893 (2004)
31. Corsello, M.A., Garg, N.K.: Synthetic chemistry fuels interdisciplinary approaches to the production of artemisinin. Nat. Prod. Rep. **32**(3), 359–366 (2015)
32. Ahmed, C.M., Burkhart, M.A., Subramaniam, P.S., Mujtaba, M.G., Johnson, H.M.: Peptide mimetics of gamma interferon possess antiviral properties against vaccinia virus and other viruses in the presence of poxvirus B8R protein. J. Virol. **79**(9), 5632–5639 (2005)
33. Avendano, C., Menendez, J.C.: Peptidomimetics in cancer chemotherapy. Clin. Transl. Oncol. **9**(9), 563–570 (2007)
34. Carmona-Ribeiro, A.M.: Biomimetic particles in drug and vaccine delivery. J. Liposome Res. **17**(3–4), 165–172 (2007)
35. Chen, S., Chen, R., He, M., Pang, R., Tan, Z., Yang, M.: Design, synthesis, and biological evaluation of novel quinoline derivatives as HIV-1 Tat-TAR interaction inhibitors. Bioorg. Med. Chem. **17**(5), 1948–1956 (2009)
36. Denap, J.C., Thomas, J.R., Musk, D.J., Hergenrother, P.J.: Combating drug-resistant bacteria: small molecule mimics of plasmid incompatibility as antiplasmid compounds. J. Am. Chem. Soc. **126**(47), 15402–15404 (2004)
37. Yang, X.X., Hu, Z.P., Duan, W., Zhu, Y.Z., Zhou, S.F.: Drug-herb interactions: eliminating toxicity with hard drug design. Curr. Pharm. Des. **12**(35), 4649–4664 (2006)
38. Khan, Y.M., Aliabbas, S., Vimal, V., Rajkumar, S.: Recent advances in medicinal plant biotechnology. Indian J. Biotechnol. **8**, 9–22 (2009)
39. Huang, H.P., Li, J.C., Huang, L.Q., Wang, D.L., Huang, P., Nie, J.S.: The application of biotechnology in medicinal plants breeding research in China. Chin. J. Integr. Med. (2015)
40. Ferraro, K.F., Kim, S.: Health benefits of religion among Black and White older adults? Race, religiosity, and C-reactive protein. Soc. Sci. Med. **120**, 92–99 (2014)
41. Adair-Rohani, H., Zukor, K., Bonjour, S., Wilburn, S., Kuesel, A.C., Hebert, R., Fletcher, E. R.: Limited electricity access in health facilities of sub-saharan Africa: a systematic review of data on electricity access, sources, and reliability. Glob. Health Sci. Pract. **1**(2), 249–261 (2010)
42. Maiyaki, M.B., Garbati, M.A.: The burden of non-communicable diseases in Nigeria; in the context of globalization. Ann. Afr. Med. **13**(1), 1–10 (2014)
43. Okechukwu, A.A., Okechukwu, O.I.: Prevalence of paediatric HIV infection in federal capital territory. Abuja. Niger J. Med. **20**(4), 409–413 (2011)
44. Uguru, N.P., Onwujekwe, O.E., Uzochukwu, B.S., Igiliegbe, G.C., Eze, S.B.: Inequities in incidence, morbidity and expenditures on prevention and treatment of malaria in southeast Nigeria. BMC Int. Health Hum. Rights **9**, 21 (2009)
45. Cloete, T.E., Nel, L.H., Theron, J.: Biotechnology in South Africa. Trends Biotechnol. **24** (12), 557–562 (2006)
46. Africa moves to strengthen biotechnology: Nature **399**(6731), 6 (1999)
47. Agbo, E.C., Agwale, S., Ezeugwu, C.O., Semete, B., Swai, H., Ikeme, A., Somiari, R.I.: Biotechnology innovation in Africa. Science **321**(5897), 1778 (2008)
48. Andanda, A.P.: Health-related biotechnology in Africa: managing the legislative and regulatory issues. Afr. J. Med. Med. Sci. **36**(Suppl), 55–61 (2007)

Biomimetics and Its Influence in Plastic and Reconstructive Surgery

Birgit Weyand and Peter Vogt

Abstract The term "biomimetics" has evolved from technical achievements based on principles found in nature. Some of its general features are found in the four areas of plastic and reconstructive surgery: reconstruction, burns, hand and aesthetic surgery. A plastic surgeon mimics concepts of nature by transplanting tissue from one to the other side or rerouting tendons or muscles to another side in order to treat local or functional defects. In contrast, with biomimetics we try to implement principles and solutions from nature in order to form or create devices, materials or technical achievements which some of them can also help to restore human tissues, body parts or body functions. This article aims to highlight interfaces between biomimetic research and principles and practice of plastic surgery.

Keywords Plastic and reconstructive surgery · Biomimetics · Biomimicry · Tissue engineering · Bionic prothesis · Biomaterial · Sensors · Sensor networking

1 Introduction and Terminology

When we talk about *biomimetics* in general, we associate the term with technological achievements which have been developed based on principles found in nature. If we just take the term *biomimetics* from its old-greek origins into its parts, namely *bios*—βίος, which means *life* and *mimesis*—μίμησις, which means imitation, we realize that this term by itself comprises something much bigger.

The term *biomimetics* was formed by Otto Schmidt in the 1950s [1]. His work was influenced by his elder brother Francis, who was working as an assistant professor in zoology at Washington University und who was studying *the molecular organization of cells and tissues with particular reference to nerve fibres* [2].

B. Weyand (✉) · P. Vogt
Department of Plastic, Aesthetic, Hand and Reconstructive Surgery, Hannover Medical School, Hannover, Germany
e-mail: weyand.birgit@mh-hannover.de

M. Israelowitz et al. (eds.), *Biomimetics and Bionic Applications with Clinical Applications*, Series in BioEngineering,
https://doi.org/10.1007/978-3-319-53214-1_3

Otto Schmidt developed an electrical device in order to mimic *the formation and propagation of impulses along nerves* [2].

After the Second World War, Otto Schmidt was able to continue his research at the University of Minnesota, where he had his first appointment in 1939. By continuing his research on devices which would mimic natural systems, he claimed a different view towards biophysics as it was commonly perceived during that time:

> Biophysics is not so much a subject matter as it is a point of view. It is an approach to problems of biological science utilizing the theory and technology of the physical sciences. Conversely, biophysics is also a biologist's approach to problems of physical science and engineering, although this aspect has largely been neglected [3].

Whilst in the English literature the term *biomimetics* or *biomimicry* is being used, in the German language the term *bionics* is being employed synonymously. However, originally the term *bionics* was taken to describe the construction of body parts with a combination of biology and electronics, which is nowadays used to cover the field of implant technology e.g. for amputees.

The term *bionics* was officially made up by Jack E. Steele during a three-day conference on "Bionics symposium: Living prototypes—the key to new technology" in 1960 at the Wright Patterson Air Force Base [4]. Jack E. Steele was a member of the US Air Force with a background in general engineering, psychiatry and neurology, and by then was working as a researcher at the 6570th Aerospace Medical Research Lab. Steele defined bionics as:

> … the science of systems which have some function copied from nature, or which represent characteristics of natural systems or their analogues. [4]

Bionics is about the systematic recognition of solutions of living nature; it distinguishes itself from purposeless nature inspiration. Bionics is based on the assumption that living nature develops optimized structures and processes through the evolutionary processes from which humans can learn.

Besides of the terms *biomimetics* and *bionics*, the term *biomimicry* has also been introduced in the field by Janine Benuys in 1997 her book: "Biomimicry: Innovation inspired by nature", where she defines it as a "… *new science that studies nature's models and then imitates or takes inspiration from these designs and processes to solve human problems*" [5].

So how does biomimetics connect to the field of Plastic and Reconstructive Surgery?

2 Principles and Practice of Plastic and Reconstructive Surgery

Plastic and Reconstructive Surgery is a relatively young surgical specialization with old roots based on four columns, namely reconstructive, aesthetic, burns and hand surgery. The main principles of plastic surgery are to restore and reconstruct form and function, often using the principle to "*replace like with like*", but by implementing donor site morbidity when "*steal from Peter to give to Paul*".

Historically, the first roots of plastic surgery can be traced back to the Edwin Smith Papyrus document, one of the oldest medical text of the ancient kingdom of Egypt around 3000–2500 B.C. containing descriptions of plastic repair of a broken nose and in India by Sushruta around 600 B.C for reconstruction of the nose by a cheek flap [6, 7]. The Sanskrit text was later translated into Arabic. In the sixteenth century, Antonius Branca and later the Italian surgeon Gaspare Tagliacozzi developed a concept of tubed and pedicled flaps for delayed transfer and reconstruction of the nose using tissue of the upper arm [8]. In the eighteenth century, the French surgeon Pierre-Joseph Desault used the term "plastic surgery" for a procedure to correct facial deformities [9].

The development of plastic surgery as a distinct surgical specialty was supported by the advances in local and general anesthesia in the late nineteenth century as well as the control of blood clotting via introduction of heparin in the early twentieth century [8, 9]. Johann Friedrich Dieffenbach was the first surgeon, who introduced ether narcotics in Germany ("Der Äther gegen den Schmerz" 1847). He and several surgical pioneers developed new methods such as free skin and fat grafting, scar release techniques such as z-plasties, local and pedicled flap surgery, facial reconstruction including nasal, ophthalmologic, ear, upper and lower lip or neck reconstruction and new advances in breast and body shaping techniques [8, 10–13].

Furthermore, the introduction of the operating microscope in the early 1920s together with microsurgical fine instrumentation and suture materials suitable for microvascular repair and neurosurgery was a milestone for the formation of the plastic surgery specialty [14].

The disabilitating injuries seen in World War I posed great challenges for surgeons of all disciplines such as Hippolyte Morestin, Harold Gillies, Erich Lexer, Varaztad Kazanijan, Hugo Ganzer, Christian Bruhn, Jaques Joseph and many others to initiate new ideas and ways to reconstruct mutilated faces, burned bodies and restore functional deficits resulting from traumatic amputations of extremities [11]. These achievements resulted in the foundation of plastic and reconstructive surgery as an autonomous surgical specialty.

Anatomical studies by Carl Manchot in 1898 about the skin arteries of the body and by Michel Simon in the 1930s using lead injection for radiographic studies of the skin and muscle arterial supply founded a basis for planning and design of tissue and muscle flaps [15, 16]. However, in the beginning of the twentieth century, the knowledge resulting from these studies was not well-known in the surgical society. Moreover, the design of local flaps for wound coverage was more or less based on random pattern with a defined base-to-length ratio rather than on a defined vascular supply.

The concepts resulting from these anatomical studies were reinvestigated in 1987 by G Ian Taylor and J H Palmer by ink injections of skin and underlying tissues [17]. The results of their research lead to the concepts of angiosomes with a specific perforator source artery supplying a three-dimensional composite tissue framework and providing an arterial roadmap for incision planning and flap design [17].

With the introduction and refinement of the microscope for use in the operating room, microsurgical free flap transfers, peripheral nerve surgery and limb replantation and revascularization after traumatic amputation became common procedures and opened a new field of reconstruction possibilities to the plastic surgeon [7, 12–14]. Further technical refinements in microsurgery opened the way to modern lymphatic surgery and perforator flap surgery [18, 19].

Nowadays, composite tissue allograft transplantation procedures such as face or limb allo-transplantation offer new reconstructive options and challenges [20]. These new methods, however, require a well-structured interdisciplinary approach for preoperative selection of suitable candidates, surgical team performance and postoperative immunological therapy and follow-up [13, 20]. Besides the demanding technical aspects of the surgical procedure, the big obstacle though is to find a suitable donor that matches the immunological characteristics of the recipient in order to be able to replace "like with like". Furthermore, the adverse long-term side effects of the required immunosuppressive therapy and the major psychological impact of these procedures on the recipients need to be considered.

3 Biomimetic Strategies for Surgical Techniques and Practices

By looking at nature's principles, several "biomimetic" strategies can be defined which already have a place in surgical research and clinical practice. We can define the following four main strategies:

1. Strategies to mimic properties of natural tissues by altering structures and functional aspects of organic or synthetic materials.
2. Strategies to mimic biological processes such as wound healing or scar formation.
3. Strategies to mimic movements or mobility which can be found in the field of bioelectrical prosthesis development.
4. Strategies to mimic complex biological processes or nano-networks for bioinspired principles for network sensor systems.

4 Strategies to Mimic Properties of Natural Tissues by Altering Structures and Functional Aspects of Organic or Synthetic Materials

The first part is implemented in the field of bioengineering and biomaterial research. Structural or functional aspects of biological tissues are being imitated by applications of synthetic or organic materials for tissue replacement and implant technology. However, so far, there is not a single material which is able to fulfill all qualities of its biological role model. Basic properties of biomaterials such as surface structure, composition, architecture, porosity and stability, structure and shape of the material might influence cell adhesion, migration, proliferation, differentiation and survival. Furthermore, functionalization of biomaterials by surface coating, or addition of molecules, proteins, growth factors or phage display technology will further influence the biological answer after implantation of the material into the body [19–24].

Research from bone tissue engineering, nerve regeneration and skin replacement have already produced a variety of medical products. These biomimetic products are then being used by surgeons in order to repair or replace tissues or body parts which are destroyed due to trauma, tumor and degenerative processes.

4.1 Bone Replacement

Artificial bone substitutes have been made from mineral composites such as hydroxyl-apatite or calcium-phosphate (which are also present in the anorganic section of bone), from ceramic or metallic material (which are inert, but may resemble bone structure by their mechanical properties), from bio-glass (which offers an osteogenic surface structure) or based on bone's organic components such as collagen in the form of synthetic collagen foams or gels or based on other fiber materials such as woven synthetic silk mats to mimic the organic tissue fraction of bone tissue [25].

The ideal material for bone replacement should support *osteoconduction*, *osteoinduction* and *osteogenesis* [26, 27]. An *osteoconductive* graft supports the ingrowth of bone from the surroundings and therefore requires an internal porous structure as well as a surface which enables bone cells to adhere and migrate. An *osteoinductive* graft supports the differentiation of osteoprogenitor cells into osteoblasts and the formation of bone extracellular matrix. This information can be provided by the material stiffness, which can be seen, e.g. in ceramics or metals such as titanium. Components of bone anorganic extracellular matrix such as hydroxyapatite or calcium phosphate can also induce differentiation of osteoprogenitor cells. Several growth factors from the group of BMP's, FGF's, VEGF's and IGF's can be osteoinductive and also support osteogenesis, but for osteogenesis active stem cells or osteoprogenitor cells are required [25, 28]. External factors such as mechanical forces, shear stresses from fluid flow or reduced oxygen levels also influence the osteogenic process as well as the dialogue between different cell types during bone formation and remodeling. Despite major achievements of past research, we still have not yet accomplished the level of being able to grow biologically full functional bone in vitro with cortical and cancellous bone components and the whole range of different types of functional cells such as osteoblasts, osteoclasts, osteocytes, fibroblasts, nerval and vascular cells. Therefore, for bone replacement surgical techniques are being used such as non-vascularized or vascularized bone transfer or the Masquelet technique to support bone formation *in loco* e.g. by providing a periostal flap coverage [25, 27, 28].

For clinical use, cell-depleted and processed frozen bone allografts (demineralized bone matrix, DBM) are also available as medical products for implantation e.g. beta-tri-calciumphoshate or hydroxyl-apatite, which can be combined with bone marrow aspirate concentrate [28, 29].

4.2 Skin Replacement

Soft tissue loss may result from various causes, which are trauma, burns, infection, arterious or venous ulcers, autoimmune disorders, or after removal of tumor lesions. Such a soft tissue defect can simply include damage of the epidermal or superficial or deep dermal layer of the skin. When a skin defect reaches the deep dermal layer, the self-regeneration potential of the skin is impaired, since it arises from stem cells of the hair bulbs located in the deep dermal layer of the skin.

Autologous split skin grafting is usually required in order to achieve coverage of the defect. When huge areas are involved, a temporary coverage with skin allograft e.g. from cadavers or even autologous transplantation of cultured keratinocyte layers might be necessary [30].

Artificial skin substitutes have been developed as bilaminated membranes with different compositions and material properties mimicking the layered organization of normal skin [30]. Examples are the combination of a collagen matrix as a dermal substitute covalently bonded to a flexible nylon fabric or with an upper outer layer of silicone rubber epidermis (e.g. Integra® Dermal Regenerative Template) [30]. Other approaches use sheets, e.g. a collagen-elastin matrix mimicking dermal composition (e.g. Matriderm®), which can be combined with autologous split skin grafts and are also slowly invaded by patients own fibroblasts, macrophages or ingrowing capillary sprouts from the surroundings [31].

Membranes made from poly-L lactide can be used in superficial partial thickness burns in order to mimic the epidermal layer to provide a temporary coverage until the wounds are re-epithelialized (e.g. Suprathel®) [32].

Further strategies combining biomaterials with immortalized cells, such as StrataGraft® which is a stratified epithelial tissue combining living dermal matrix with fibroblasts overlaid by normal immortalized human keratinocytes or Apligraf® which is a bovine collagen matrix containing a combination of human keratinocytes and fibroblasts derived from neonatal foreskin [32]. Transplantation kits for spray application of keratinocyte suspensions together with fibrin glue are also commercially available (ReCell™) [30].

Tissue engineering strategies serve as well-known examples, which intend to build or grow functional tissues or organs mainly outside the human body for later implantation.

An example is the expansion of autologous keratinocytes in vitro and the generation of cultured epidermal autografts (CEA's) for skin transplantation in severe burn victims, where more than 50 or 60% of the body surface is affected [30]. More recent approaches, which are currently still in the experimental stage, have used three-dimensional printing of complex skin substitutes with keratinocytes, fibroblasts and other cells (endothelial cells, melanocytes, neural cells etc.) in a fibrin, hydrogel or collagen matrix [31]. Other research strategies have described tissue-engineered skin substitutes formed by laser printing, cultivation of skin composites in an air-fluid interphase in order to establish a corneated epidermal layer or cultivation of vascularized skin derivates [30, 31, 33].

New approaches use immunocompetent 3D skin models in order to better understand the role of the immune system in skin biology and tissue engineering [34].

4.3 Nerve Replacement

Biomimetic technologies for neural scaffolds use material components, mimic inner architecture and surface structures of anatomical organization of original nerves and use biological modifications such as growth factors or cell additives in order to achieve a close resembling to its biological counterparts [35].

Nerve repair requires tension-free suture which can be achieved only in defects of 5 mm or less without the need of transplants. For nerve defects larger than 5 mm size, biologic or synthetic nerve guide conduits as scaffold for neural regeneration are being used. The golden standard nowadays are nerve autografts, which have their limitation and disadvantages due to second side of surgery, donor side morbidity, mismatch of diameter, size and shape and limited availability. Biological alternatives for small nerve gaps of less than 3 cm size are conduits from small vessels like peripheral veins, which might be also filled with small muscle fibres in order to promote nerve growth and conduction [36–38]. Use of allografts (or xenografts) requires immunosuppressive therapy and therefore has considerable limitations from medical and economical aspects. Various bio-adsorbable nerve guide conduits based on collagen type I, poly-glycolic acid (PGA) or poly (DL-lactide-*co*-e-caprolactone) (PLCL) have been approved by the US Food and Drug Administration, which vary in respect to their inner diameter, porosity, rigidity and biocompatibility and degeneration rates [35]. Regarding their different properties of the nerve conduits, direct comparison studies between the different materials are not always feasible, since e.g. the variance between the diameter of the nerve conduit and the diameter of the study nerve chosen will affect nerve regeneration [39]. Best results with nerve conduits have been seen in nerve defects smaller than 3 cm [39].

Another option for nerve replacement is the use of acellularized nerve allografts (ANA). ANA's are nerve allografts, were the immunogenic components such as the highly antigenic Swann cells are removed by decellularizing techniques such as detergent processing, radiation, freeze-thawing cycles or cold preservation [39, 40]. The preserved endoneuronal architecture together with collagen and laminin within the basal lamina guides cell migration and nerve fiber sprouting [39]. Commercially available and FDA-approved acellularized nerve transplants (Avance™ Nerve Grafts) have been shown superiority to collagen conduits (Integra NeuraGen®) and comparable results to autografts for small diameters (1–2 mm and short gap (<3 cm) [41].

In experimental settings in animals and as healing approach in humans in compliance with ethical standards (approved by the ethics committee of Hannover Medical School) very long nerve grafts (>10 cm) from saphenous veins filled with spider silk have been shown to be immunologically well tolerated from the recipient side. Results of reconstructed 6 cm nerve gaps in sheep showed excellent functional regeneration comparable to autologous nerve grafts [42].

Despite advances in tissue engineering and generation of artificial nerve grafts, the main prerequisites for successful recovery of peripheral nerve injury are still the process of active axonal sprouting and the preservation of functional motoric end plates during the repair process.

5 Strategies to Mimic Biological Processes Such as Wound Healing or Scar Formation

Whilst skin substitutes can survive the first days after transplantation solely by diffusion from the underlying wound bed, other tissues such as bone require immediate connection to the blood circulation and also an internal preformed capillary bed for sufficient oxygen supply. Tissue engineering strategies in order to improve vascularization of the host involve the implantation of cell-free autologous or allograft connective tissue structural components with preformed capillary networks in order to facilitate vessel ingrowth [43, 44]. Engineered tissue matrices with free or attached growth factors or hormones for time and place-controlled-staged release for angiogenesis and vascular ingrowth have already entered clinical testing for special indications [43, 44]. Further strategies imply transplantation of vessels into the host bed or transplantation of endothelial cells which have been isolated and cultivated from the host beforehand [43, 44]. Moreover, application of autologous fat stem cells into chronic problematic wounds, fistulae, radio-dermatitis or instable scar tissues have demonstrated impressive effects on wound healing, scar tissue remodeling or local vascularity by exhibiting paracrine effects by growth factors, hormones, cell-cell interactions and cell-extracellular-matrix interaction [45, 46].

For wound therapies, many approaches mimic closely the staged wound repair process of our body by providing an enzymatic or surgical preparation of the wound bed, e.g. the application of plated-rich plasma [47], or the use of polymers or a connective tissue layer to give a structure e.g. for capillary ingrowth [33, 48, 49], or to provide a closed environment to support the reepithelization process [31, 50].

6 Strategies to Mimic and Support Movements or Sensibility Which Can Be Found in the Field of Bioelectrical Prosthesis Development

Loss of functionality of limbs can be caused by an amputation or also by traumatic peripheral nerve injuries, by spinal cord or cerebral injury or by ischemic, immunologic or vascular events. When nerve pathways get injured, nerve reflexes and circuits are being interrupted and distorted, leading to flaccid or spastic paresis or paralysis of the affected limb, loss of sensibility and chronic pain. The internal repair capacity of the central and peripheral nervous system is limited, resulting in lasting deficits. Glial scar formation from astrocytes, oligodendrocytes and Schwann cells produces factors that inhibit neurite outgrowth, remyelination and axonal repair [51].

Furthermore, when injury causes partial or complete transection of peripheral nerves, surgery cannot restore the electrical transmission potential of the nerve itself. Surgery can only reconstruct the outer nervous sheaths, which is mandatory for guided axonal regeneration, by direct epineural sutures or insertion of nerve grafts for lager defects.

Secondary surgical options in order to reconstruct muscle function are neurotization procedures by transfer of intact motoric nerves onto a distal motoric nerve stump leading towards a paralyzed muscle which may allow muscle regeneration as long as functional motor end plates are preserved. Exemplary, in total brachial plexus injuries, the biceps brachii muscle for elbow flexion can be reinnervated by transfer of several motoric branches of the intercostal nerves to the distal musculocutaneus nerve. If muscle paralysis was prolonged or local muscle tissue had been severely damaged, a free microvascular functionalized muscle transfer with connection to local vascular and nerve bundles is another option. For instance, a free gracilis muscle can be connected to the microvascular and functioning neural network of the recipient side.

Another reconstruction option can be achieved by means of muscle or tendon transfers: Lost mobility may be regained and activities of daily living improved by giving up power and strength of donor muscle functions however. A well-known procedure is the transfer of wrist flexors into wrist and finger extensors in radial palsy in order to improve function of the hand.

When parts of limbs are completely lost, a prosthetic device can substitute form and function to a certain extent. The use and development of prostheses has a long history.

The first models were simply walking aids or cosmetic prothesis: Famous examples are a wooden prosthesis of a big toe dated back to 1000–600 B.C which had been found in a sarcophagus of ancient Egypt [52], and a lower limb prosthesis from the time of the Roman Empire around 300 B.C. which had been located in Capua.

Prothesis development further proceeded with mechanical or body powered devices, where positioning was adjusted by means of harnesses, lines or straps.

From medieval times, the articulated hand of Ambrosius Paré in 1579 or the iron hand of Goetz from Berlichingen with two different designs around 1510 and 1530 are two examples [52]. However, these prostheses had to be adjusted and aligned by the healthy other hand in order to allow some grip functions [52, 53]. Around 1818, the German dentist Peter Baliff, designed a prosthetic device, where muscular forces from the shoulder girdle and trunk were transferred via leather straps into the terminal device attached to the arm stump, which later inspired Giuliano Vanghetti and Ferdinand Sauerbruch to prepare amputation stumps by kineplastic operations to allow transfer of muscle and tendon forces directly onto the prosthesis in order to allow active movements [53, 54]. The Bowden cable body-powered prosthesis from 1948 was another example for a mechanical prosthetic device which allowed the user to control a two-pronged hook via preserved shoulder and body movements by variable tensions of cables.

With technical advances of the twentieth century, electrically-powered, myo-electric prosthesis were created, which are controlled by sensors amplifying surface electro-myographical potentials from contracting residual muscles in the prosthetic socket. In Germany, the first myoelectric prosthesis was created by Reinhold Reiter, a physics student at Munich University, and presented 1948 at a fair trade in Hannover [55]. Another myoelectric arm prosthesis designed by the Russian Alexander Kobrinski in 1960 went into commercial production and was also tested in British and Canadian institutes [53]. Obviously, the weight of the batteries for power supply was a challenge for the first models. Furthermore, the initiation and proceeding of movement commands into the mechanical reaction chain were slow, and changing sensor positions due to sweating or moving distorted electromyographical signals [53, 55].

Modern approaches use different interfaces to link the human body with neuroprothesis and hybrid bionic systems [56]. The connecting interfaces between prosthetic devices and the nervous system can be based on chemical, mechanical, magnetic or electrical sensors and can be invasive versus non-invasive [56].

Targeted muscle reinnervation (TMR) was introduced by Kuiken and describes the transfer (rerouting) of remaining peripheral nerves (e.g. median and ulnar nerve) after amputation to preserved muscles, e.g. the chest muscles. The amplified neuronal signals from the muscle amplifiers can be then transduced through e.g. transdermal sensors which pick up the muscular signals from underneath and allow control of the prosthesis functions by their specific determined nerves [52, 57].

For bionic limbs, the connection between the body and the prosthetic device requires a stable mechanical interface, a dynamical interface to support movements and an electrical interface to connect and communicate with the muscle and nervous system. Nowadays there are new compound materials with better mechanical properties which allow change from soft and flexible to hard and durable stabilizers to connect between the amputation stump, skin and the prosthesis [58, 59]. Another option are the so-called endo-exoprothesis where the artificial limb can be directly fixed with the bone via osseointegrated titan pins, providing therefore maximal stability and endogenous proprioception of the prosthetic limb [53, 60].

The electrical interface can be achieved by sensors which record signals from the muscular surface, or directly inside muscles, or directly within or around nerves by implanted sieve or cuff electrodes. Long-term stability and reliability as well as possible damage to the structures have to be considered, especially with implanted nerve sensors.

The interface electrodes have to meet the standard requirements of compatibility between a technological and a biological system, where from the technological sight the interface device is a bi-directional transducer to record bioelectrical signals from the body (muscles, nerves etc.), whereas from the biological perception, the interface resembles a foreign body [56].

Non-invasive electrodes are surface electrodes usually applied to the skin which can record signals from muscles, central nervous system or from the heart. They also can be used to activate peripheral nerve tracts for sensation, chronic pain suppression transcutaneous electrical neural stimulation (TENS) or even as a functional electrical stimulation (FES) to correct foot drop [56]. In experimental settings, non-electrical interfaces working by magnetoneurography to detect nerve action potentials or by acustic myography of muscle vibrations during voluntary or provoked contraction have been tested [56, 61]. Muscle electrodes are either placed on the muscle surface or within the muscle fibers and can be used for stimulation or recording. Whereas most of these electrode systems function by direct electrical coupling, there is also one interesting microstimulator device which can be activated by coupling to a wireless control system by an externally generated radiofrequency magnetic field [62]. Extra neural electrodes can be placed epineural or in form of "cuff"-electrodes around the nervous sheat.

Other electronical devices may improve lost functions in central, spinal cord or peripheral nerve lesions: Deep brain stimulation uses "brain pacemakers" in order to apply controlled electronic impulses to specific targets in brain areas ("brain nuclei") which are an accepted treatment in Parkinson's disesases, essential tremor or dystonia/ataxia. Spinal cord stimulation may help patients with chronic back pain by triggering and suppression of nociceptive pathways in the spinal cord. A peripheral paralysis might be improved by implantation of functional electrostimulation systems close to peripheral nerves such as the peroneus nerve. The effectivity however depends on the degree of residual function which is intended to be enforced.

7 Strategies to Mimic Complex Biological Processes or Nano-Networks for Bioinspired Principles for Network Sensor Systems

Nature has been an inspiration for computer scientists and engineers to mimic nature principles to design network sensor systems which can adapt to various changing conditions, self-organize, adjust scaling and provide robust an durable operation for long-term survival. Many biological systems have an internal dynamic which is guided by a small number of simple rules, do not require a central controlling head and allow synchronization, task allocation, resource management, and efficient communication with adaption to environmental changes [63]. Examples of nature-inspired principles which have been used for the design of network sensor systems are swarm intelligence, particularly ant colony optimization and the particle swarms optimization algorithms, natural time synchronization, artificial immune system and intercellular information exchange [63, 64]. These natural principles are not only used for sensor development for medicinal devices, but also used for other computer and robotics applications in tracking, filtering, selecting, signaling or detection of invading elements, to name a few examples.

8 Conclusion

In the field of plastic and reconstructive surgery, biomimetic principles can help to replace parts which the surgeon otherwise has to borrow from other body parts or locations. Biomimetic strategies may revolutionize new approaches in reconstructive surgery and medicine, such as tissue engineering, artificial intelligence, sensor development or material sciences: Tissue or body parts or their functions are newly re-created outside the body in order to be implanted at a later time point. Further challenges include the connection with the peripheral nervous systems and the human brain functions as a central driving motor via biomimetic sensor systems as well as the implementation of neural networks in the control module. A cross-disciplinary working relationship between material scientists, engineers, physicists, chemists, mathematicians and medical doctors has been essential for current and will be for future achievements.

References

1. Vincent, J.F.V.: Biomimetics—a review. **223**(Part H), 919–39 (2009)
2. Harkness, J.M.: In appreciation a lifetime of connections: Otto Herbert Schmitt, 1913–1998. Phy. Perspect. **4**(4)456–490 (2002)
3. Schmitt, F.O.: The never-ceasing search. Philadelphia, vol. 188 (1990)
4. Robinette, J.C.: Living Prototypes—The Key to New Technology: Bionics Symposium ; 13–14–15 September 1960. Wright-Patterson Air Force Base, Ohio (WADD technical report; Bd. 60) (1960)
5. Benyus, J.M.: Biomimicry : innovation inspired by nature. William Morrow 308 S (1997)
6. Shiffman, M.A., Di Giuseppe, A.: Herausgeber. Cosmetic surgery: art and techniques. Springer, Heidelberg, New York 1192 S (2013)
7. Singh, V.: Sushruta: the father of surgery. Natl. J. Maxillofac Surg. **8**(1), 1–3 (2017)
8. Tennenhaus, M., Lukascz, L., Ogawa, R., Rennekampff, H.: A brief historical review of flaps and burn reconstruction. Wounds **20**(7), 214–218 (2008)

9. Thompson, L.: Plastic Surgery (Health and Medical Issues Today). Greenwood Pub Group (2011)
10. Chick, L.R.: Brief history and biology of skin grafting. Ann. Plast. Surg. **21**(4), 358–365 (1988)
11. Thomas, R.L., Fries, A., Hodgkinson, D.: Plastic Surgery Pioneers of the Central Powers in the Great War (2018)
12. Wallace, A.F.: History of plastic surgery: the early development of pedicle flaps. J. Royal Soc. Plast. Surg. **71**, 834–838 (1978)
13. Whitaker, I.S., Karoo, R.O., Spyrou, G.M., Fenton, O.M.: The birth of plastic surgery: the story of nasal reconstruction from the Edwin Smith Papyrus to the twenty-first century. Plast. Reconstr. Surg. **120**, 327–336 (2007)
14. Tamai, S.: History of microsurgery. Plast. Reconstr. Surg. **124**(6 Suppl), e282–e294 (2009)
15. Manchot, C.: The cutaneous arteries of the human body. [Hautarterien des menschlichen Körpers]. Springer Science and Business Media, LCC (1983)
16. Salmon, M., Taylor, G.I., Tempest, M.N.: Arteries of the Skin. Churchill Livingstone 174 S (1988)
17. Taylor, G.I., Palmer, J.H.: The vascular territories (angiosomes) of the body: experimental study and clinical applications. Br. J. Plast. Surg. **40**, 113–141 (1987)
18. Clodius, L., Piller, N., Casley-Smith, J.: The problems of lymphatic microsurgery for lymphedema. Lymphology **14**(2), 69–76 (1981)
19. Becker, C., Vasile, J., Levine, J., Batista, B., Studinger, R., Chen, C., u. a.: Microlymphatic surgery for the treatment of iatrogenic lymphedema. Clin Plast Surg. **39**(4), 385–98 (2012)
20. Diaz-Siso, J., Bueno, E., Sisk, G., Marty, F., Pomahac, B., Tullius, S.: Vascularized composite tissue allotransplantation–state of the art. Clin. Transpl. **27**(3), 330–337 (2013)
21. Jaggessar, A., Shahali, H., Mathew, A., Yarlagadda, P.K.D.V.: Bio-mimicking nano and micro-structured surface fabrication for antibacterial properties in medical implants. J. Nanobiotechnol. **15**(1) (2017) Verfügbar unter: http://jnanobiotechnology.biomedcentral.com/articles/10.1186/s12951-017-0306-1
22. Sailaja, G.S., Ramesh, P., Vellappally, S., Anil, S., Varma, H.K.: Biomimetic approaches with smart interfaces for bone regeneration. J. Biomed. Sci. **23**(1), 77 (2016)
23. Green, D.W., Watson, G.S., Watson, J.A., Lee, D.-J., Lee, J.-M., Jung, H.-S.: Diversification and enrichment of clinical biomaterials inspired by Darwinian evolution. Acta Biomater. **42**, 33–45 (2016)
24. Martins, I.M., Reis, R.L., Azevedo, H.S.: Phage display technology in biomaterials engineering: progress and opportunities for applications in regenerative medicine. ACS Chem. Biol. **11**(11), 2962–80 (2016)
25. Brett, E., Flacco, J., Blackshear, C., Longaker, M.T., Wan, D.C.: Biomimetics of bone implants: the regenerative road. BioRes. Open Access. **6**(1), 1–6 (2017)
26. Myerhoff, C., Archdeacon, M.: Autogenous bone graft: donor sites and techniques. J. Bone Joint Surg. Am. **93**, 2227–36
27. Buck, D.W., Dumanian, G.A.: Bone biology and physiology: Part II. Clinical correlates. Plast. Reconstr. Surg. **129**, 950e (2012)
28. Verboket, R., Leiblein, M., Seebach, C., Nau, C., Janko. M., Bellen, M., u. a.: Autologous cell-based therapy for treatment of large bone defects: from bench to bedside. Eur. J. Trauma. Emerg. Surg. Off Publ. Eur. Trauma Soc. **44**(5), 649–65 (2018)
29. Duan, R., Barbieri, D., Luo, X., Weng, J., Bao, C., de Bruijn, J.D., u. a.: Variation of the bone forming ability with the physicochemical properties of calcium phosphate bone substitutes. Biomater. Sci. **6**(1), 136–45 (2017)
30. Wood, F.M.: Skin regeneration: the complexities of translation into clinical practise. Int. J. Biochem. Cell Biol. **56**, 133–140 (2014)
31. Girard, D., Laverdet, B., Buhé, V., Trouillas, M., Ghazi, K., Alexaline, M., u. a.: Biotechnological management of skin burn injuries: challenges and perspectives in wound healing and sensory recovery. Tissue Eng. Part B **23**(1), 59–82 (2017)

32. Stone, L.I.R., Natesan, S., Kowalczewski, C., Mangum, L., Clay, N., Clohessy, R., u. a.: Advancements in regenerative strategies through the continuum of burn care. Front. Pharmacol. **9**, 672 (2018)
33. Michael, S., Sorg, H., Peck, C.-T., Koch, L., Deiwick, A., Chichkov. B., u. a.: Tissue engineered skin substitutes created by laser-assisted bioprinting form skin-like structures in the dorsal skin fold chamber in mice. PloS One **8**(3), e57741 (2013)
34. Pupovac, A., Senturk, B., Griffoni, C., Maniura-Weber, K., Rottmar, M., McArthur, S.L.: Toward immunocompetent 3D skin models. Adv. Healthc. Mater. **7**(12), 1701405 (2018)
35. Du, J., Chen, H., Qing, L., Yang, X., Jia, X.: Biomimetic neural scaffolds: a crucial step towards optimal peripheral nerve regeneration. Biomater. Sci. **6**(6), 1299–1311 (2018)
36. Battiston, B., Titolo, P., Ciclamini, D., Panero, B.: Peripheral nerve defects. Hand Clin. **33**(3), 545–550 (2017)
37. Brunelli, G.A., Battiston, B., Vigasio, A., Brunelli, G., Marocolo, D.: Bridging nerve defects with combined skeletal muscle and vein conduits. Microsurgery **14**(4), 247–251 (1993)
38. Geuna, S., Tos, P., Titolo, P., Ciclamini, D., Beningo, T., Battiston, B.: Update on nerve repair by biological tubulization. J. Brachial. Plex. Peripher. Nerve. Inj. **9**(1), 3 (2014)
39. Mackinnon, S.E., Yee, A.: Nerve Surgery: MediaCenter.thieme.com includes videos online. Thieme, New York 627 S (2015)
40. Moore, A.M., MacEwan, M., Santosa, K.B., Chenard, K.E., Ray, W.Z., Hunter, D.A., u. a.: Acellular nerve allografts in peripheral nerve regeneration: a comparative study. Muscle Nerve. **44**(2), 221–34 (2011)
41. Whitlock, E.L., Tuffaha, S.H., Luciano, J.P., Yan, Y., Hunter, D.A., Magill, C.K., u. a.: Processed allografts and type I collagen conduits for repair of peripheral nerve gaps. Muscle Nerve. **39**(6), 787–99 (2009)
42. Radtke, C., Allmeling, C., Waldmann, K.-H., Reimers, K., Thies, K., Schenk, H.C., u. a.: Spider silk constructs enhance axonal regeneration and remyelination in long nerve defects in sheep. Egles C, Herausgeber. PLoS ONE. **6**(2), e16990 (2011)
43. Johnson, P.C., Mikos, A.G., Fisher, J.P., Jansen, J.A.: Strategic directions in tissue engineering. Tissue Eng. **13**(12), 2827–2837 (2007)
44. Lovett, M., Lee, K., Edwards, A., Kaplan, D.L.: Vascularization strategies for tissue engineering. Tissue Eng. Part B Rev. **15**(3), 353–370 (2009)
45. Cherubino, M., Rubin, J.P., Miljkovic, N., Kelmendi-Doko, A., Marra, K.G.: Adipose-derived stem cells for wound healing applications. Ann. Plast. Surg. **66**(2), 210–215 (2011)
46. Kallmeyer, K., André-Lévigne, D., Baquié, M., Krause, K.-H., Pepper, M.S., Pittet-Cuénod, B., u. a.: Fate of systemically and locally administered adipose-derived mesenchymal stromal cells and their effect on wound healing. Stem. Cells Transl. Med. (2019)
47. Fernandez-Moure, J.S., Van Eps, J.L., Cabrera, F.J., Barbosa, Z., Medrano del Rosal, G., Weiner, B.K., u. a.: Platelet-rich plasma: a biomimetic approach to enhancement of surgical wound healing. J. Surg. Res. **207**, 33–44 (2017)
48. Kapoor, S., Kundu, S.C.: Silk protein-based hydrogels: promising advanced materials for biomedical applications. Acta Biomater. **31**, 17–32 (2016)
49. Trask, R.S., Williams, H.R., Bond, I.P.: Self-healing polymer composites: mimicking nature to enhance performance. Bioinspir. Biomim. **2**(1), 1–9 (2007)
50. Cerqueira, M.T., Pirraco, R.P., Martins, A.R., Santos, T.C., Reis, R.L., Marques, A.P.: Cell sheet technology-driven re-epithelialization and neovascularization of skin wounds. Acta Biomater. **10**(7), 3145–3155 (2014)
51. Bradbury, E.J., McMahon, S.B.: Spinal cord repair strategies: why do they work? Nat. Rev. Neurosci. **7**(8), 644–653 (2006)
52. Bumbaširević, M., Lesic, A., Palibrk, T., Milovanovic, D., Zoka, M., Kravić-Stevović, T., u. a.: The current state of bionic limbs from the surgeon's viewpoint. EFORT Open Rev. **5**(2), 65–72 (2020)
53. Zuo, K.J., Olson, J.L.: The evolution of functional hand replacement: from iron prostheses to hand transplantation. Plast. Surg. Oakv. Ont. **22**(1), 44–51 (2014)

54. Tropea, P., Mazzoni, A., Micera, S., Corbo, M.: Giuliano Vanghetti and the innovation of "cineplastic operations". Neurology. **89**(15), 1627–32 (2017)
55. McLean, L., Scott, R.N.: The early history of myoelectric control of prosthetic limbs (1945–1970). In: Muzumdar, A. (eds) Herausgeber. Powered Upper Limb Prostheses [Internet]. Springer Berlin Heidelberg, Berlin, Heidelberg (2004) [zitiert 23. October 2020]. S. 1–15. Verfügbar unter: http://link.springer.com/10.1007/978-3-642-18812-1_1
56. Navarro, X., Krueger, T.B., Lago, N., Micera, S., Stieglitz, T., Dario, P.: A critical review of interfaces with the peripheral nervous system for the control of neuroprostheses and hybrid bionic systems. J. Peripher. Nerv. Syst. **10**(3), 229–258 (2005)
57. Kuiken, T.A.: Targeted muscle reinnervation for real-time myoelectric control of multifunction artificial arms. JAMA. **301**(6), 619 (2009)
58. Haines, C.S., Li, N., Spinks, G.M., Aliev, A.E., Di, J., Baughman, R.H.: New twist on artificial muscles. Proc. Natl. Acad. Sci. USA **113**(42), 11709–11716 (2016)
59. Palza, H., Zapata, P.A., Angulo-Pineda, C., Electroactive smart polymers for biomedical applications. Mater. Basel. Switz. **12**(2) (2019)
60. Ortiz-Catalan, M., Hakansson, B., Branemark, R.: An osseointegrated human-machine gateway for long-term sensory feedback and motor control of artificial limbs. Sci. Transl. Med. **6**(257), 257re6–257re6 (2014)
61. Barry, D.T., Leonard, J.A., Gitter, A.J., Ball, R.D.: Acoustic myography as a control signal for an externally powered prosthesis. Arch. Phys. Med. Rehabil. **67**(4), 267–269 (1986)
62. Loeb, G.E., Peck, R.A., Moore, W.H., Hood, K.: BION system for distributed neural prosthetic interfaces. Med. Eng. Phys. **23**(1), 9–18 (2001)
63. Hylsberg Jacobsen, R., Zhang, Q., Skjødeberg Toftegaard, T.: Bioinspired principles for large-scale networked sensor systems: an overview. Sensors. **11**(4), 4137–51 (2011)
64. Dressler, F., Akan, O.B.: A survey on bio-inspired networking. Comput. Netw. **54**(6), 881–900 (2010)

Biomimetics Models

Torsional Magnetic Angle for *Magnetospirillum gryphiswaldense*

Sarah Strauß, Meir Israelowitz, Birgit Weyand, Robert Müller, Henkel Thomas, Dirk Schüler, René Uebe, Syed W. H. Rizvi, Christoph Gille, Herbert P. von Schroeder, Kerstin Reimers, and Peter Vogt

Abstract In the present study we tested magnetosomes isolated from the magnetic bacterium *Magnetospirillum gryphiswaldense* for their magnetic qualities when embedded in three different environmental surroundings: (1) medium (Phosphate buffer saline PBS), (2) gelatin gel matrix and as (3) powder. (3) is used as base line (standard control). We studied the magnetic hysteresis, thermal degradation in relation to theoretical results obtained with natural magnetosomes, by using an indirect analysis with biomimetic principles.

Keywords *Magnetospirillum gryphiswaldense* · Magnetic dipole moment of the crystal · Magnetic field · Torque

Abbreviations

MTB	Magnetotactic bacteria
M(H)	Hysteresis loops M (H)
FF	Solution of standard phosphate buffered saline (PBS)
P	Dry powder denoted as P
FF-G	Immobilized in a gelatin gel matrix denoted

S. Strauß (✉) · B. Weyand · K. Reimers · P. Vogt
Department of Plastic, Aesthetic, Hand and Reconstructive Surgery, Hannover Medical School, Hannover, Germany
e-mail: strauss.sarah@mh-hannover.de

M. Israelowitz · S. W. H. Rizvi · C. Gille · H. P. von Schroeder
Biomimetics Technologies Inc, Toronto, Canada

R. Müller · H. Thomas
Leibniz Institute of Photonic Technology, Jena, Germany

D. Schüler · R. Uebe
University of Bayreuth, Bayreuth, Germany

H. P. von Schroeder
University of Toronto, Toronto, Canada

M. Israelowitz et al. (eds.), *Biomimetics and Bionic Applications with Clinical Applications*, Series in BioEngineering,
https://doi.org/10.1007/978-3-319-53214-1_4

Hc Coercivity
Mr Magnetic remanence
Ms Magnetic saturation

1 Introduction

Biological systems are equipped with highly specialized sensors to respond to (environmental) changes in their natural niches [1]. For example, insects use sensors for orientation to achieve specific biological activities [2–5]. Another example are magnetotactic bacteria (MTB) which orient and migrate along geomagnetic field lines to find growth favoring an- or microoxic zones within aquatic sediments [6, 7]. Therefore, MTB intracellularly biomineralize membrane-enclosed magnetite (Fe_3O_4) or greigite (Fe_3S_4) nanoparticles, termed magnetosomes. Magnetite and greigite are inverse spinel iron minerals in which anions of the crystal lattice are packed in a close cubic array with eight occupied tetrahedral sites and sixteen occupied octahedral sites with 32-anions per magnetite unit cell [8]. The electrons of the outermost *p* orbital anions interact in three dimensions with the iron cations (positively charged) in a fashion that aligns the unpaired electrons of the tetrahedral cations opposite to the unpaired electrons of the octahedral cations [9, 12]. The number of unpaired electrons in the two structures (octahedral and the tetrahedral) create ferrimagnetism that results in a sensitivity of MTB to magnetic fields [10]. The responsiveness of MTBs to the weak geomagnetic field, however, is only possible due to the chain-like arrangement of individual magnetosomes within the cell (Fig. 1).

This organization induces the magnetic moments of individual single domain state magnetite particles to align with each other and, thus, increases the magnetic

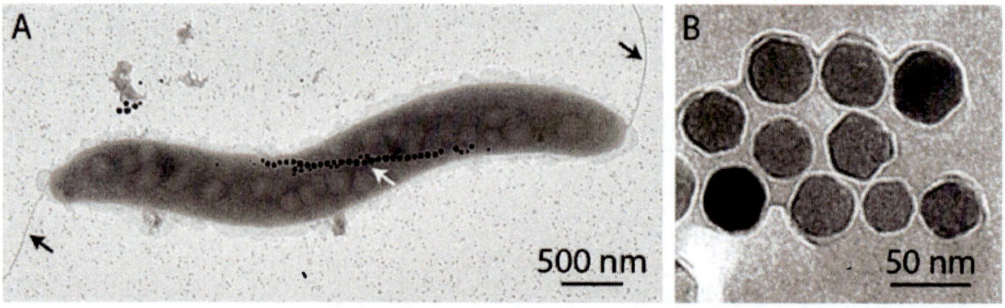

Fig. 1 **a** Transmission electron micrograph of the freshwater α-proteobacterium *Magnetospirillum gryphiswaldense* which produces up to approximately 100 magnetite particles per cell with an average diameter of 35 nm that are aligned as a single chain along the longitudinal cell axis. Black arrows, flagellar filaments; white arrow, position of magnetosome chain. **b** Transmission electron micrograph of isolated magnetosomes from *M. gryphiswaldense* still enclosed by the magnetosome membrane

energy of the magnetosome chain to that of the sum of all particle magnetic moments within the chain. Thereby, magnetosomes provide an evolutionary advantage due to the ability of efficient navigation within zones of sharp chemical gradients by simplifying a potential three dimensional search to a single dimension.

A fundamental problem for any sensors, whether biological or human instrumentation, is noise (in our context: thermal motion of electrons (Johnson Current Noise)) [11]. In order to overcome the noise for instrumentation, the system needs to be cooled down to 77 K [12–14]. A biological system is able to overcome noise by using multiple detectors to average the signal ratio as in our case the bacteria by using multiple magnetosomes [16]. The study at hand presents the underlying mathematical-physical principles.

2 Software

For the simulation of the model we used Mathematica 10.0 [17] because the power of the software allowed work with symbols for consideration of magnetic fields.

3 Methods

3.1 Detector

Quasistatic magnetic properties were tested using MicroMag™ 3900 vibrating sample magnetometer (VSM) (Princeton Measurement Corp., Ohio, USA). Hysteresis loops M(H), with a maximum field up to 14,000 Oe (1 Oe = 1/4π kA/m) and remanence curves were measured. Quasistatic hysteresis losses were calculated from minor loops and the switching field distribution from the initial remanence curves [18]. Coercivity values and remanence of hysteresis loops at the saturation measured at different directions were examined for texture effects. All measurements were performed at room temperature (22 °C).

3.2 Magnetospirillum gryphiswaldense

Magnetospirillum gryphiswaldense was grown microaerobically in flask standard medium [19] for 24 h at 25 °C as described earlier [20]. Cells were harvested by centrifugation (10,500 xg, 20 min, 4 °C) and washed twice with ice cold wash buffer (20 mM Hepes pH 7.4, 5 mM EDTA). Cell pellets were stored at −80 °C until use. Magnetosome isolation and purification with minor modifications was performed according to the protocol of Rosenfeldt et al. [21].

4 Model

Most of the crystals from *Magnetospirillum gryphiswaldense* adopt an octahedral {1 1 1} morphology, but crystals with cubic {1 0 0} or dodecahedral {1 1 0} morphology also occur [22, 23]. The crystal size is 35 nm; the width {1 1 1} is 4.9 Å, with a length {2 $\bar{2}$ 0} of 3.0 Å, [24].

5 Results

Measurements were performed with several magnetic field strength to simulate the magnetic gradients of the magnetic field of the earth along which the bacteria orientate. Magnetosome samples were divided into three groups, one in solution of standard phosphate buffered saline (PBS) denoted as FF; the second a dry powder denoted as P; and thirdly FF immobilized in a gelatin gel matrix denoted as FF-G. All samples were exposed to a magnetic field to measure the effects on the magnetic orientation in the crystals in the magnetic field. The magnetization curve M(H) (magnetitization due to the magnetic field (H)) [25, 26] was normalized to the sample mass. From saturation magnetization values of a pure dried magnetic phase (Fe_3O_4 or γ-Fe_2O_3 (maghemite) the concentration of iron oxides in a fluid sample can be determined as shown later on. In our case the Ms-values are diminished by the mass fraction of diamagnetic organic material adhered to the iron oxide.

The magnetic coercivity Hc (M = 0), (Hc: coercivity is a measure of the ability of a ferromagnetic material to withstand an external magnetic field without becoming demagnetized) [27], remanence (H = 0) (hysteresis) [28] and Mr/Ms (Mr: magnetic remanence, Ms magnetisation) [29], depends on the particle size, shape, mobility, interaction and orientation [30]. Figure 2 shows magnetization curves M(H) with different maximum field strength (see colors) for a powder (Fig. 2a), i.e. statistical orientation of the particles. Figure 2b and c show such curves of the gelatin samples after a texturing in a magnetic field, measured parallel (2b) and perpendicular (2c) to the texture axis. The saturation magnetization of the powder sample is 16.3 Am^2/kg (i.e. the dry sample contains $\approx$ 18% Fe_3O_4 (assuming bulk material properties) or 25% of γ-Fe_2O_4 with the typical properties (pure iron oxide without organic coating). An aging effect was observed (decrease to 12% of first measurement), probably due to oxidation of Fe_3O_4 to γ-Fe_2O_3 [31]. The textured gel-samples show a soft- and a magnetically harder phase. The properties depend on the direction of the measurement.

For a better comparison magnetization curves of the different samples (at the same maximum field strength) are shown in Fig. 2.

Figure 3a shows field dependent magnetization curves of particles with different mobility that were normalized to the magnetization values at maximum field strength and were measured before (a) and gel samples after (b) the texturing in a

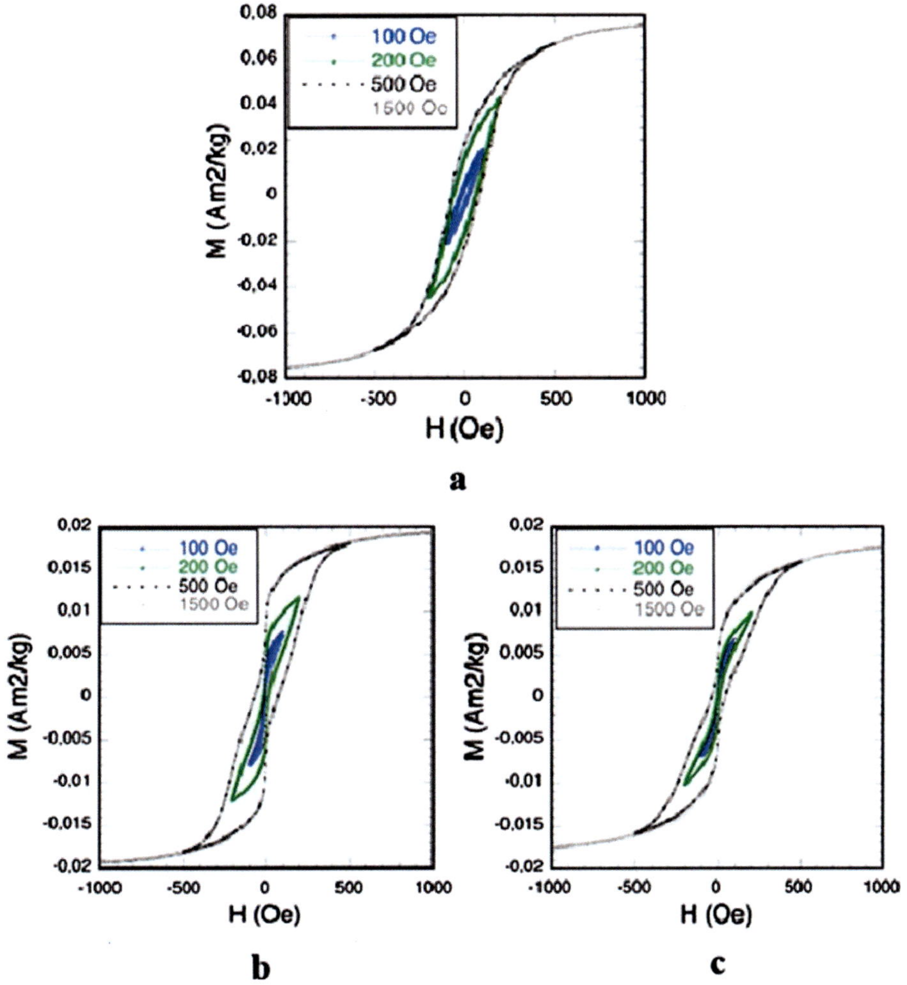

Fig. 2 a–c Field dependent magnetization curves with different maximum field strength (see colors) of a powder sample (2a) with statistical particle orientation and textured gel samples parallel and perpendicular to the texturing axis

static field depending on the direction, i.e. magnetic field during the measurement was parallel ("para") or perpendicular ("perp") to the texturing axis, affected by the field.

Figure 3a shows particles with no mobility and (b) shows a magnified excerpt with an expanded *x*-axis, where Hc, Mr of FF > 0 and (c) shows the particle mobility affected by the field.

Figure 4a reveals a remanence curve normalized to the saturation remanence Mr (H)/Mr(H_{max}) and Fig. 4b is the switching field distribution calculated from the remanence curves, revealing the fraction of particles which switch irreversibly at a certain field strength and is proportional to the product of particle volume and an

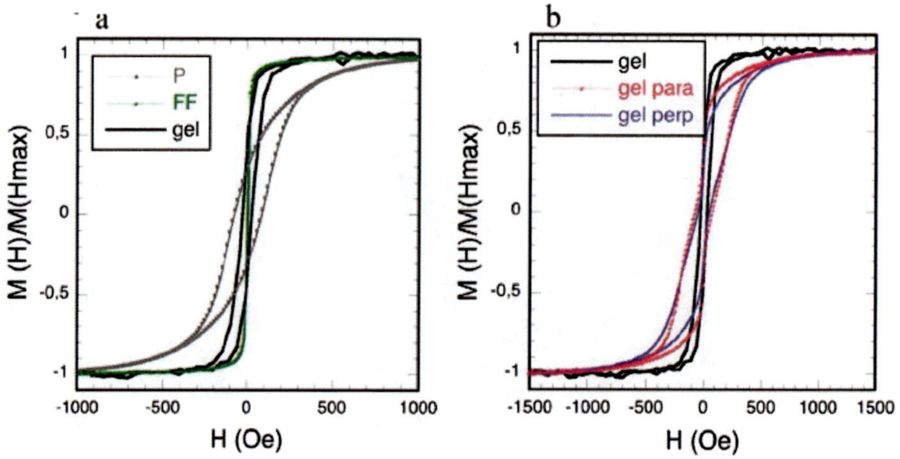

Fig. 3 a–b, field dependent magnetization curves (H_{max} = 1500 Oe) of powder, fluid and gel sample (left) and of gel sample before texturing and after, depending on direction of measurement

effective magnetic anisotropy (that is small and independent from the size according to the TEM images), with more sensitivity to the sizes of the particles than M(H) curves. Figure 4a FF reveals a shoulder at ca. 700 Oe that cannot be explained by free movable soft magnetic particles and might be caused by agglomeration/texturing.

Figure 5 shows the Hc and Mr/Ms in gelatin before texturing (blue) and after (green) depending on the direction of measurement (0°: texturing is parallel to measurement axis).

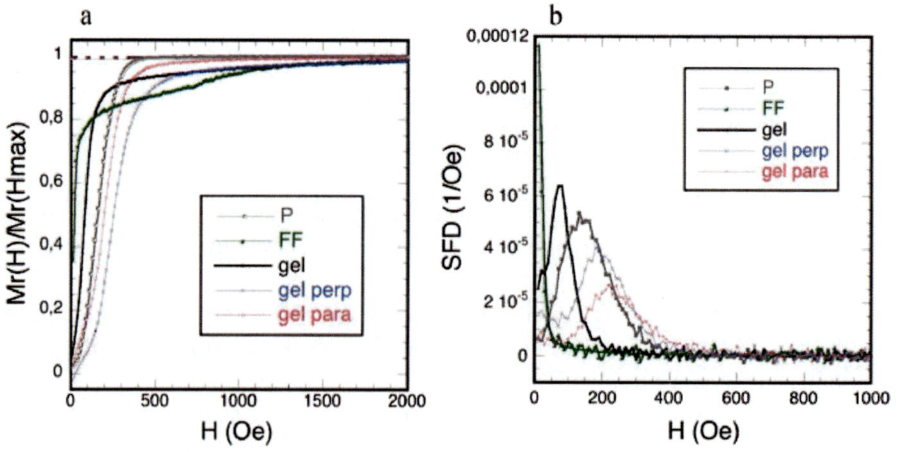

Fig. 4 a–b Remanence curves (left) and calculated switching field distributions (right)

Fig. 5 Shows hysteresis parameters Hc and Mr/Ms (blue gel) depending on the direction of measurement before and after texturing of a gel sample in a magnetic field

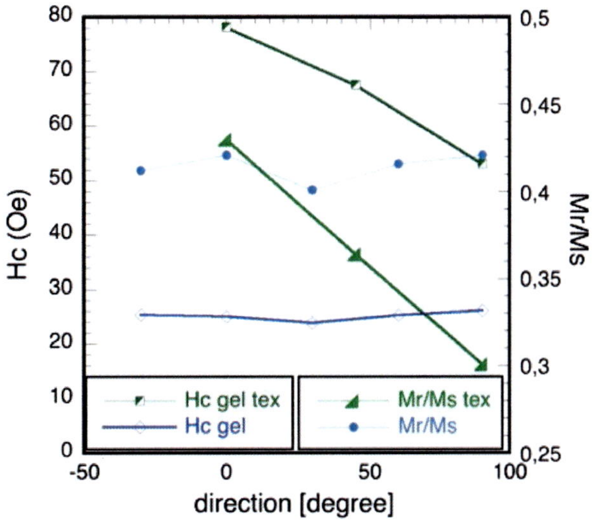

6 Discussion

In order to minimize the energy within large ferromagnetic minerals multiple domains develop. Each domain has a net magnetic moment and each domain is aligned differently from each other and, thus, reduces the net magnetic moment of the whole crystal [32–36]. Each crystal is different from each other, as we show size, texture, shape affect the magnetic moment [33, 35, 37–44]. For small crystals there is only a single magnetic moment so there is a strong resistance to a new orientation [45, 46]. Thus, to change the orientation there is a need of change in temperature or external force such as a magnetic field [47, 48].

6.1 Magnetic Dipole Moment of the Crystal

To understand the magnetosome we can consider it as nano-crystal, where, the magnetic xy is above ~35 nm (i.e. we consider single domain particle) [49] and monopole does not apply [50–59], therefore single crystals can be considered as a single magnet. The properties of the electrical field crystal with the sizes: a = 4.9 Å, b = 3.0 Å, c = 100 nm [60, 61] the simulation follow the Miller principles, especially the law of rational indices and the electrical field is given by

$$E = {}^{\sigma}\!/_{2\varepsilon_0} \quad (1)$$

where $\sigma = {}^{q}\!/_{A}$ and $q = {}^{I}\!/_{v}$ is the charge in a static magnetic field, where I is the current and v is time take the charge to move a distances at time, ε_0 is electrical permittivity. Using the electrical field the magnetic field can be calculated for a single domain.

$$v = 1.5962 \times 10^{-21}\left({}^{m}/_{s}\right) * \text{electron}/\text{electron and } q = 1.7544 \times 10^{-8}\,{}^{C}/_{\text{electron}}$$

$$E = 9.9102\,{}^{V}/_{m}$$

The electrical field is given by:

$$\Phi_E = \oiint_E Eds = {}^{q}/_{\varepsilon_0} = 2.5759x10^{-27}\,{}^{Nm^2}/_{C} \quad (2)$$

The model considered is consistent with a magnet on a wire, where the direction of the field is comparable to the gradient.

6.2 *Magnetic Field*

$$B = {}^{\mu_0 I}/_{2\pi r} = 1.692x10^{-5}\,{}^{Ns}/_{Cm} \quad (3)$$

6.3 *Torque*

$$\tau = IL^2 B \sin\theta = \mu B \sin\theta = 4x10^{-34} Nm \quad (4)$$

where τ is proportional to the current I, the flux density B and the loop area L. Work torsional angle Fig. 6, knowing the shape of the crystals and assume and field conditions is constant for the model. Figure 7 show the sizes versus the intensity of the magnetic field and Fig. 8 show the density distribution of the torque [62, 63].

The experimental data static hysteresis losses per cycle normalize to the specific saturation value of dried powder sample (fluid and gelatin sample are not normalize to the particle mass).

Hysteresis losses (Fig. 9) at saturation field are in the same order of magnitude as sample values in the literature [64], and conform to the theoretical magnetic dipole, torsional magnetic angle, which depends on the texture. However the high values in the fluid sample in the low range field are unusual. A similar behavior was found for particles in a fluid with pure superparamagnetic particles, i.e. particles that show not (only) Neel relaxation [65].

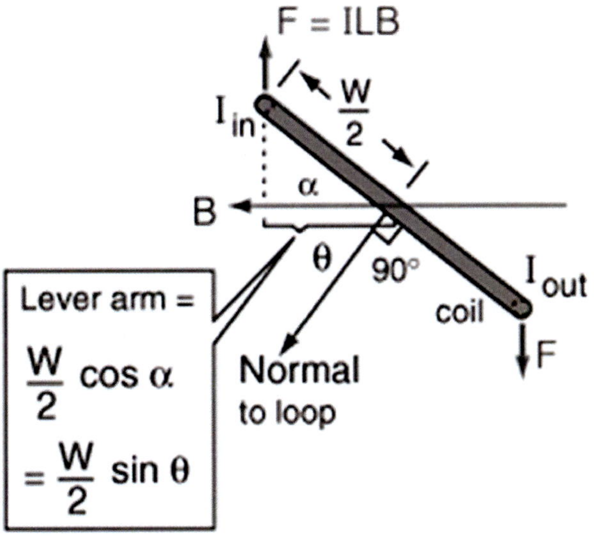

Fig. 6 Shows the geometry of the torque, the direction of the forces

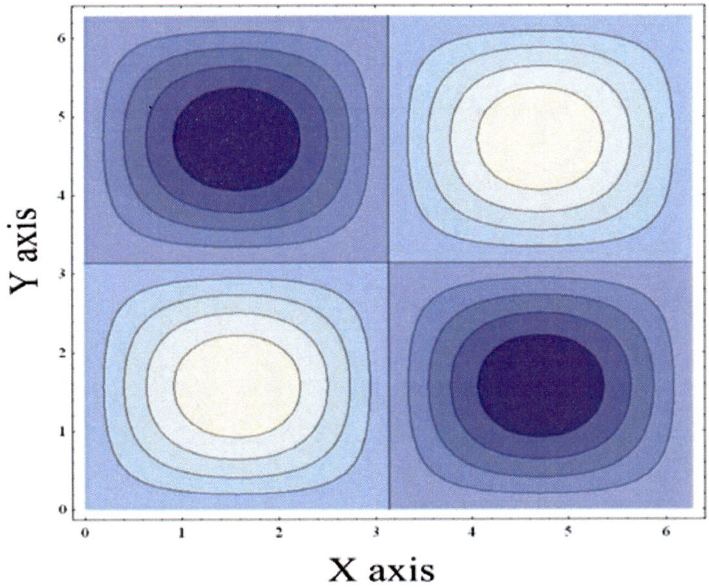

Fig. 7 Shows the poles of single magnetosomes, the size in nm (the sizes are in the xy coordinates) and the intensity of the torque Nm, since the measure of the torque, the intensity is in the poles (the intensity is given by the colour intensity, [0 is no torque (white) to 1 Newtons per meter for the higher intensity (purple)])

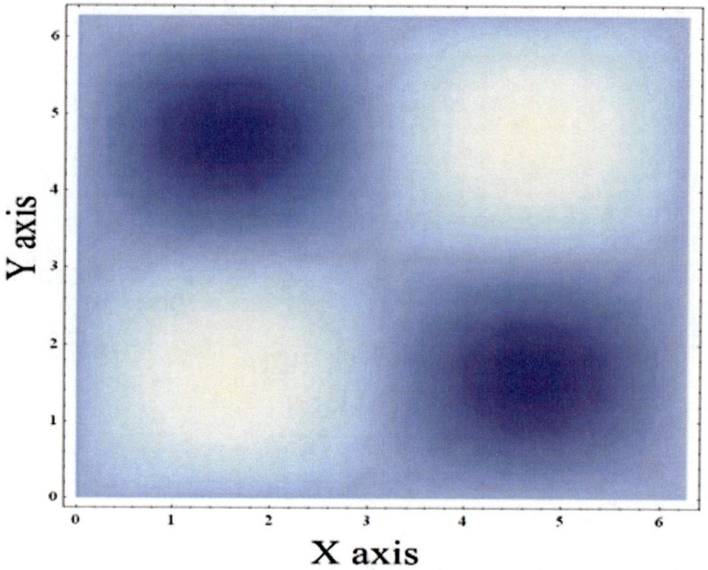

Fig. 8 Shows the magnetic density field distribution of the torque (the sizes in nm, xy-axis). The torque depends on the anisotropic shape of the sizes of the e magnetosomes [density is Teslar, where 0 (white) is not intensity to 1 where is the maximal intensity 1 Teslar (purple)]

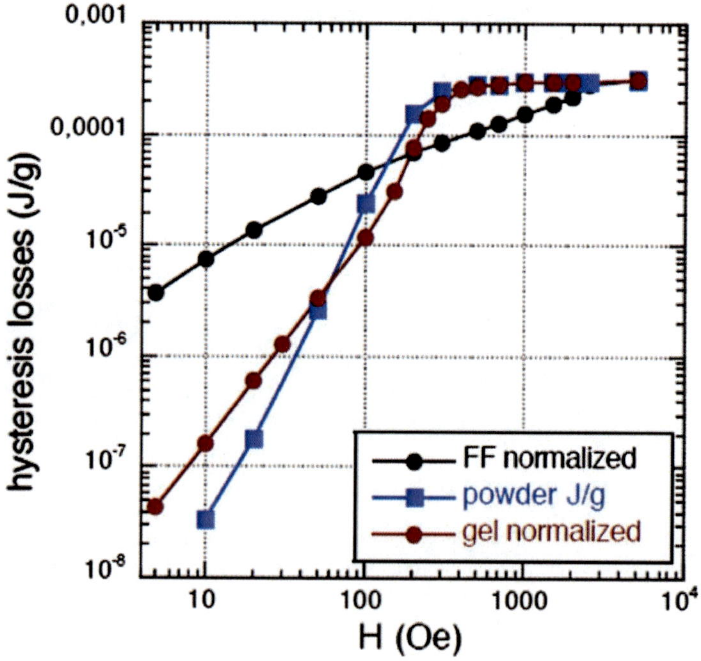

Fig. 9 Static hysteresis losses per magnetization cycle normalized to the specific saturation value of the dried powder sample (fluid and gelatin sample are NOT normalized to particle mass)

7 Conclusion

For some applications (e.g. medical applications, mineral detection) bacterial magnetite offers several advantages compared to chemically synthesized magnetite, since bacterial magnetites have a consistent shape and a narrow size distribution within the single magnetic domain range and therefore allow a more refine magnetic detection reducing the noise signaling. Furthermore the magnetosome envelope has membrane coating lipids and proteins that might allow easy coupling of bioactive substances to its surface [66].

The prerequisite for any large scale commercial application of magnetosomes is mass cultivation of MTBs or the introduction and expression of the genes responsible for magnetosome synthesis into different bacteria *(e.g., E. coli)* that can be grown relatively easy to large yields [67, 68].

The study at hand shows that it is feasible to design a system to mimic magnetotactic behavior/properties. Measurements of the magnetic field and the torsional angle, showed individual characteristics of magnetosomes, explaining how *Magnetospirillum gryphiswaldense* overcomes the thermal noise by electron movement in the crystal. As the system works at room temperature it might be translated to other applications (e.g. in medical sensor systems) which suffer at the moment from too high noise signaling [5].

References

1. Wolken, J.J.: Light Detectors Photoreceptors, and Imaging Systems in Nature. Oxford University Press (1995)
2. Israelowitz, M., Rizvi, S.W., von Schroeder, H.P.: J. Lumin. **126**, 149 (2007)
3. Israelowitz, M., Kwon, K.A., Rizvi, S.W., Gille, C., von Schroeder, H.P.: J. Bionic Eng. **8**, 129 (2011)
4. Israelowitz, M., Weyand, B., Leiterer, C., Munoz, V., Martinez-Tomas, C., Herraiz-Llacer, M., Slowik, I., Beleites, C., Fritzsche, W., Krafft, C., Henkel, T., Reuter, M., Rizvi, S., Gille, C., Reimers, K., Vogt, P., von Schroeder, H.P.: New J. Sci. (2014)
5. Israelowitz, M., Rizvi, S.W., Gille, C., Holm, C., von Schroeder, H.P.: (n.d.)
6. Bellini, S.: Su Di Un Particolare Comportamento Di Batteri D'acqua Dolce Instituto Di Microbiologia dell'Universita Di Pavia. Tipografia Popolare (1963)
7. Blakemore, R.: Science **190** (1975)
8. Klein, C.: Manual of Mineral Science 22nd Eds. John Wiley and Sons, New York (2002)
9. Banerjee, S., Moskowitz, B.: In: Magn. Biominer. Magnetoreception Org. New Biomagn, pp. 17–42. Plenum Press, New York (1985)
10. Favrie, D., Schüler, D.: Chem. Rev. **108**, 4875 (2008)
11. Lee, S.K., Romalis, M.V.: J. Appl. Phys. **103**, 084904 (2008)
12. van der Ziel, A.: Noise in Solid State Devices and Circuits. Wiley, New York, NY, USA (1986)
13. Frieden, B.R.: Probability, Statistical Optics and Data Testing. Springer, New York, NY, USA (1991)
14. Marasco, P.L. Dereniak, P.L.: In: SPIE, pp. 363–378 (1993)

15. Dereniak, D.L., Boreman, G.D.: Infrared Detectors and Systems. John Wiley & Sons, New York (1996)
16. Pfeiffer, D., Schüler, D.: Quantifying the benefit of a dedicated "Magnetoskeleton" in bacterial magnetotaxis by live-cell motility tracking and soft agar swimming assay. Appl. Environ. Microbiol. **86**(3), (2020). https://doi.org/10.1128/AEM.01976-19
17. Wolfran, S.: Mathematica: Ein System für Mathematik auf dem Computer. Addison-Wesley Publishing Company (1994)
18. Pfeiffer, H.: Phys. Sataus. Solidi. A **118**, 295 (1990)
19. Heyen, U., Schüler, D.: Appl. Microbiol. Biotechnol. **61**, 536 (n.d.)
20. Barber-Zucker, S., Uebe, R., Davidov, G., Navon, Y., Sherf, D., Chill, J.H., Kass, I., Bitton, R., Schüler, D., Zarivach, R.: Disease-homologous mutation in the cation diffusion facilitator protein MamM causes single-domain structural loss and signifies its importance. Sci. Rep. **6**, 31933 (2016). https://doi.org/10.1038/srep31933
21. Rosenfeldt, S., Mickoleit, F., Jörke, C., Clement, J.H., Markert, S., Jérôme, V., Schwarzinger, S., Freitag, R., Schüler, D., Uebe, R., Schenk, A.S.: Towards standardized purification of bacterial magnetic nanoparticles for future in vivo applications. Acta. Biomater. **120**, 293–303 (2021). https://doi.org/10.1016/j.actbio.2020.07.042
22. Bazylinski, D.: Chem. Geol. **135**, 191 (n.d.)
23. Devouard, B., Posfal, P., Huan, X., Bazylinski, D., Frankel, R., Buseck, P.: Am. Mineral. **83**, 1387 (1998)
24. Klein, C., Dutrow, B.: Mineral Science, 23nd edn. Wiley (2007)
25. Preisach, F.: Z. Für Phys. **94**, 277 (1933)
26. Liorzou, F., Phelps, B., Atherton, D.L.: IEEE Trans. Magn. **36**, 418 (2000)
27. Akhter, M.A., Mapps, D.J., Ma Tan, Q.Y., Petford-Long, A., Doole, R., Tan, M., Doole, P.L.: J. Appl. Phys. **81**, 4122 (n.d.)
28. Jiles, D.C., Atherton, D.L.: J. Magn. Magn. Mater. **61**, 48 (1986)
29. Stoner, E.C., Wohlfarth, E.P.: R. Soc. Lond. A **240**, 599 (n.d.)
30. Zhang, Y., Fei, C., Wang, R., Yang, G., Xiong, R., Shi, J.: Hydrothermal method. Send J. Nanosci. Nanotechnol. **10**, 6395 (2010)
31. Maki, D., Homburg, J.A., Brosowske, D.A.: Archeol. Prospect. **13**, 207 (2006)
32. Cullity, C., Graham, D.: Introduction to Magnetic Materials, 2nd edn. Wiley-IEEE, New York (2008)
33. Jiles, D.: Introduction to Magnetism and Magnetic Materials. Chapman & Hall, London (1998)
34. Weiss, P.: Comptes Rendus **143**, 1136 (1906)
35. Carey, R., Issac, E.D.: Magnetic Domains and Techniques for Their Observation. The English University Press Ltd, London (1996)
36. Feynman, R.P., Leighton, R.B., Sands, M.: The Feynman Lectures on Physics. California Inst. of Technology, US (1963)
37. Bedanta, S., Kleemann, W.: J. Phys. **42**, 013001 (2009)
38. Bean, C.P., Livingston, J.D.: J. Appl. Phys. **30**, 120S (1959)
39. Guimarães, A.P.: Principles of Nanomagnetism. Springer, Berlin/Heidelberg, Germany (2009)
40. Frenkel, J., Doefman, J.: Nature **126**, 2714 (n.d.)
41. Mørup, S., Hansen, M.F., Frandsen, C.: Comprehensive Nanoscience and Technology Andrews. In: Scholes, D., Wiederrecht, G., (eds). Elsevier, Amsterdam, The Netherlands (2011)
42. Bertotti, G.: Hysterisis in Magnetism: For Physicists, Materials Scientists, and Engineers. Academic Press-Elsevier, Waltham, MA, USA (1998)
43. Gubin, S.P., Kossharov, Y.A., Khomutov, G.B., Yurkov, G.Y.: Russ. Chem. Rev. **74**, 489 (2005)
44. Issa, B., Obaidat, I.M., Albiss, B., Haik, Y.: Int. J. Moelcular Sci. **14**, 21266 (2003)

45. Dobrynin, A.N., Ievlev, D.N., Temst, K., Lievens, P., Margueritat, J., Gonzalo, J., Alfonso, C. N., Zhou, S.Q., Vantomme, A., Piscopiello, E., van Tendeloo, G.: Appl. Phys. Lett. **87**, 012501 (2005)
46. Kodama, R.H.: J. Magn. Mater. **200**, 359 (n.d.)
47. Strbak, O., Kopcansky, P., Frollo, I.: Measument Sci. Rev. **11**, 85 (2011)
48. Barnejee, S.K., Moskowitz, B.M.: Plenum Publ. Corp. 17 (1985)
49. Sone, K., Naganuma, H., Takamichi, I., Takashi, M., Okuruma, S.: Sci. Rep. **5**, 9348 (2015)
50. Preskill, J.: Phys. Rev. Lett. **43**, 1365 (1979)
51. Giblin, S.R., Bramwell, S.T., Holdsworth, P.C.W., Prabhakaran, D., Terry, I.: Nat. Phys. **7**, (2011)
52. Price, P.B., Shirk, E.K., Osborne, W.Z., Pinsky, L.S.: Phys. Rev. Lett. **35**, 487 (1975)
53. Cabrera, B.: Phys. Rev. Lett. **48**, 1378 (1982)
54. Preskill, J.: Ann. Rev. Nucl. Part Sci. **34**, 461 (1984)
55. Moulin, F.: Nuovo. Cimento. **116**, 869 (2001)
56. Castelnovo, C., Moessner, R., Sondhi, S.L.: Nature **451**, 42 (2008)
57. Ray, M.W., Ruokokoski, E., Kandel, S., Möttönen, S., Hall, D.S.: Nature **505**, 657 (2014)
58. Dirac, P.: Proc. Roy. Soc. Lond. A **133**, 60 (1931)
59. Wolfgang, R.: Am. J. Phys. **57**, 993 (1989)
60. Battle, X., Labarta, A.: J. Phys. Part D **35**, 3069 (2001)
61. Ho, C.-H., Lai, C.-H.: IEEE Trans. Magn. **42**, 3069 (2006)
62. Jackson, J.D.: Classical Electrodynamics 3rd edn. Wiley (1998)
63. Krauss, J.D.: Electronmagnetics 4th edn. McGraw-Hill Series in Electrical Engineering (1991)
64. Nedkov, I., Vanderberghe, R.E., Tailhades, P.: Nanosized Magnetite for Biomedical Applications, in Nanomedicine Design of Particles, Sensors, Motors, Implants, Robots, and Devices. In: Mark, J., Schulz, V., Shanov, N., Yeoheung Y. (eds). Artech Houses (2009)
65. Müller, R., Dutz, S., Hergt, R., Schmidt, C.H., Steinmetz, H., Zeisberger, Gawalek, Z.: J. Magn. Magn. Mater. **310**, 2399 (2007)
66. Mickoleit, F., Lanzloth, C., Schüler, D.: A versatile toolkit for controllable and highly selective multifunctionalization of bacterial magnetic nanoparticles. Small **16**(16), e1906922. https://doi.org/10.1002/smll.201906922
67. Zhang, Y., Zhang, X., Jiang, W., Li, Y., Li, J.: Appl. Env. Microbiol. **77**, 5851 (2011)
68. Kolinko, I., Lohße, A., Borg, S., Raschdorf, O., Jogler, C., Tu, Q., Pósfai, M., Tompa, E., Plitzko, J.M., Brachmann, A., Wanner, G., Müller, R., Zhang, Y., Schüler, D.: Nat. Nanotechnol. Lett. **9**, 193 (2014)

Spider Silk as Biomaterial for Medical Applications and Tissue Engineering

Malte Fließ and Sarah Strauß

Abstract A spider is capable of producing 7 different subtypes of threads, each of which fulfilling highly specialized, biological functions that has been developed over 500 million years of evolution. Due to its mechanical and biological characteristics spider silk (especially the dragline) is in focus of various scientific fields. This chapter gives a general view of potential medical applications of spider silk and tissue engineering approaches.

Keywords *Nephila edulis* · Silk reeling · Biocompatibility · Adipose-derived stroma cells

1 Introduction

Due to its remarkable properties such as high elasticity, exceptional tensile strength and break resistance, spider silk has been closely examined for diverse applications in a variety of subject areas [1, 2] in recent years. Especially the *dragline,* the exceptionally durable while easily obtainable thread for the web's outer rim and spokes, nowadays is well characterized and described in regards of its tertiary and quaternary protein structure as well as its hierarchical superstructure of different layers [3, 4]. A spider is capable of producing 7 different subtypes of threads, each of which fulfilling highly specialized, biological functions, that have been developed over 500 million years of evolution.

Considering history it becomes evident that spider silk has been used by mankind not only in everyday life (i.e. fishery by Solomon Islands natives [5]) but also

In memoriam of Professor Dr. Kerstin Reimers (*16.01.1970 †23.12.2015), head of the experimental department of the department of plastic, aesthetic, hand and reconstructive surgery, at Hannover Medical School.

M. Fließ · S. Strauß (✉)
Department of Plastic, Aesthetic, Hand and Reconstructive Surgery, Hannover Medical School, Hannover, Germany
e-mail: strauss.sarah@mh-hannover.de

M. Israelowitz et al. (eds.), *Biomimetics and Bionic Applications with Clinical Applications*, Series in BioEngineering,
https://doi.org/10.1007/978-3-319-53214-1_5

for medical applications over millennia. Ancient Greek and Roman physicians, such as *Claudios Galen of Pergamon*, are known to have treated miscellaneous wounds with spider silk and its hemostatic properties were used until the middle Ages [6]. As a result of its high biocompatibility, spider silk recommends itself as genuine biomaterial. During *Christianization* of the occident, the medical use of spider silk was banished as "*sorcery*" and thus a lot of knowledge was lost.

Historically, the rediscovery of spider silk is led by material scientists focusing on its unique material properties, but a prerequisite for a deeper examination of the material was the ability to collect spider silk in consistent quality as well as in sufficient amounts. Under these circumstances, the golden orb-weaver spider (*Nephila* spec.) seems to be particularly eligible. The female spider weaves webs with diameters up to 2 m, is capable of producing up to 500 m of dragline in one piece and is widely common in tropic and subtopic areas. With suitable environmental conditions, the reproduction rate of *Nephila edulis* spiders (Figs. 1 and 2) is fairly stable.

As early as 1796, the Spanish Jesuit abbot *Raimondo Maria de Termeyer* experimented with spider silk and developed equipment to fixate a single spider and collect its silk on a bobbin manually [7].

Based on this device Richard Work designed a machine which is a predecessor of today's widely used electrified winding devices [8].

2 Silk Reeling

Nowadays sophisticated winding devices have significantly facilitated the reeling process by means of variable coiling speed and digital thread length measurement. The process is neither deadly nor painful for the animal, as the dragline production is a physiological process. Figure 3 illustrates the reeling set-up.

On a soft foam block the animal is positioned on its back and covered with fine gauze. The gauze is then fixed to the foam block with needles between the spider's legs, leaving its anterior-medial silk gland (*Glandula ampullate majoris*) uncovered. In general, a piece of dragline hangs out of the silk gland (Fig. 4) which is then attached to the spool of the winding device. To start the reeling process, the device is switched on, setting the spool into rotation. The animals are conditioned to the process easily by limiting the procedure to 15 min and rewarding the cooperation with food.

3 Biocompatibility of Spider Silk

As mentioned earlier, its biocompatibility is a fundamental prerequisite for the usage of spider silk in the biomedical field. Though its application over millennia might recommend itself, the certification and licensing process for medical products

Fig. 1 Female *Nephila edulis* in typical position

demands valid data. Prior to data collected on its potential *cyto-* and *hemotoxicity* (by an accredited test laboratory in accordance with **DIN EN ISO 17025**), initial studies focused on the compatibility of spider silk and murine fibroblast cell lines [9]. The results showed no direct toxic effects of spider silk on the used cells [unpublished data]. The in vitro data are confirmed and supplemented by in vivo tests on rats, sheep and pigs [10–13]. In fact, the dragline silk caused less inflammatory response than other biomaterials [10, 11]. It even seems that spider silk is degraded completely over different periods of time depending on the

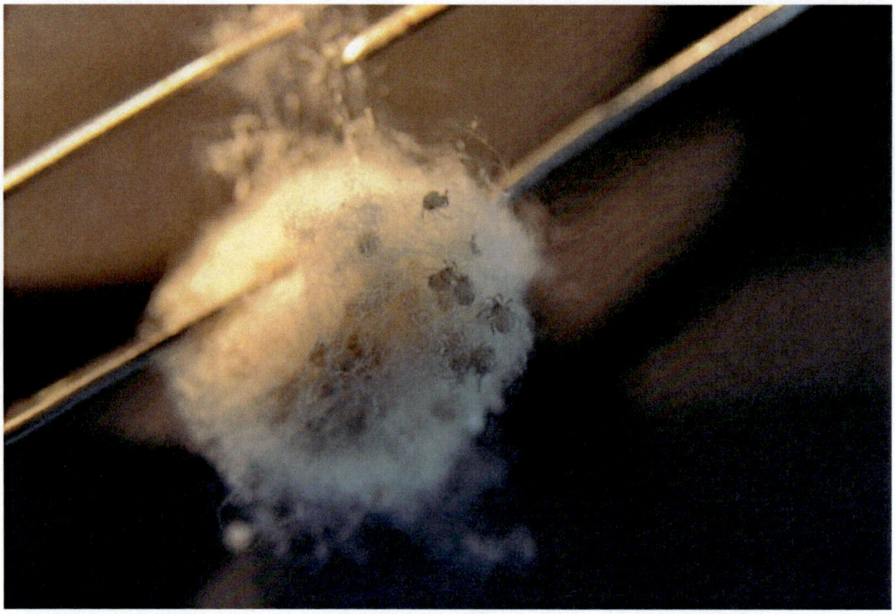

Fig. 2 Few days old spiderlings (*N. edulis*) in cocoon

implantation site. While a participation of giant cells is postulated, the exact mechanism of degradation inside the tissue still remains unclear [11, 13].

4 Implantation

The fascia surrogate utilizing rats shows infiltration of giant cells in chronological sequence with the degradation of spider silk tissue. In this experiment spider silk threads were twisted into miniature cord and then weaved into net fabric. After implantation the net caused minimal immunogenicity compared to material widely used clinical application (Surgisis®, a biological, porcine, acellular collagen matrix of small intestine mucosa and Ultrapro®, an artificial matrix made of polypropylene and poliglecapron) while maintaining constant strength and elasticity until degradation. First signs of decomposition could be observed after 4 weeks, while the spider silk could no longer be determined histologically after 4 months [10]. The surrounding tissue formed sturdy scar tissue, supporting the defect permanently [10]. This could predestine spider silk as an alternative for synthetic, non-degradable materials which achieve sustainable stabilization but could cause constant irritation of the surrounding tissue leading to chronic pain in the long run. Following the plastic-surgical approach of replacing "*like with like*", the use of spider silk aims for regeneration rather than repair.

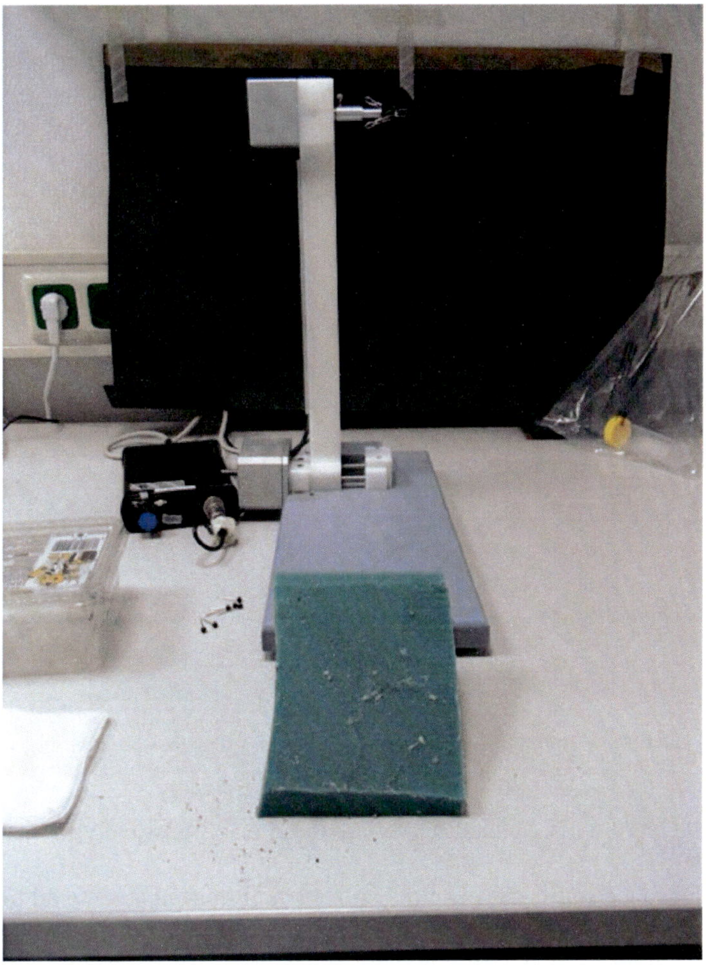

Fig. 3 Set up for reeling of dragline silk

5 Results

Closed miniature ropes made of spider silk represent another possible application in the biomedical context. Today, *polyamide* (*nylon*) suture material represents the gold standard for peripheral neurosurgery. A study focusing on morphology and mechanical properties indicated that suture material made of twisted spider silk threads of comparable elasticity and diameter possess 2.5-fold the tensile strength of polyamide thread. In addition, spider silk suture material showed no fatigue under cyclic load and thus proves to be superior to other standard material such as *Polypropylen* [14, 15]. The degradation of spider silk, unlike absorbable sutures like *PLA* (*polyactid*), causes no alteration of the pH-value, which especially favors

Fig. 4 Ventral sight on the abdomen with silk glads of a female *N. edulis*

the regeneration of peripheral nerves, as these react sensitively to changes of this parameter [16].

This advantage also relates to the utilization of transplant material to reconstruct peripheral nerve defects of critical size. By means of spider silk constructs, *Nervus Ischiadicus* defects of sheep up to 6 cm were treated successfully [8]. The construct consisted of a spider silk rope inside a *decellularized* vein. The ends of the spider silk were then sutured to the *epineuron* of the remaining nerve stumps. The evaluation 10 month *post operationem* showed complete morphological regeneration and *remyelination*. Electro *neurographical* tests confirmed these results. As observed earlier with the fascia surrogate, the spider silk could not be determined histologically during the evaluation (after 10 month).

Three-dimensional cross woven skin substitute scaffolds can be produced by coiling spider silk around small dental wire frames (Fig. 5). The scaffold can be effectively sterilized by steam-pressure process and applied in cell culture or in vivo. Artificial skin is produced with cross woven scaffolds of mesh sizes between 10 and 100 μm and colonized with dermal fibroblast cells. By migration and proliferation, the cells fill the silk mesh from the periphery to the center. After approx. one week of proliferation, the top of the scaffold becomes colonized with keratinocytes. Then, after another proliferation phase, differentiation of *kerationcytes* is induced by supplementation with 0.4 μg/ml Ascorbat-2-phosphate and cultivation at an air-liquid interface. After 3 weeks of incubation the tissue shows a

Fig. 5 Dental wire frame with cross woven spider silk for cell culture applications

typical skin-structure with elongated fibroblasts of the dermis and flattened cells of the epidermis [17]. In contrast to in vivo application the spider silk is not degraded in vitro, leaving the matrix intact and thus maintaining its stability when cultivation is completed. The resilience of the artificial tissue reduces the risk of damage during clinical transplantation significantly. This is of utmost importance as, up to now, artificial skin has failed to reach the mechanical characteristics of either *autologous*, *allogeneic* or *xenogenic* transplants.

Another successful application of spider silk in tissue engineering is the experimental generation of musculoskeletal tissue. In particular, adipose-derived stroma cells (ASCs) are a promising starting point, as they feature the stem cell characteristics comparable to mesenchymal stroma cells from bone marrow. Under sufficient stimulation ASCs can be differentiated into *osteogenic*, *chondrogenic*, *tenogenic* or *adipogenic* form, rendering primary cell isolation from bone marrow with its often low yields unnecessary.

Three-dimensional scaffolds with appropriate mechanical properties are a necessity to generate functional tissue in short preparation and cultivation time, allowing earliest in vivo implantation. Especially a tear resistance and tensile strength comparable to tendon and ligaments is mandatory, which is given and even surpassed by spider silk. Hence scaffolds made of this material are ideal for cultivation and differentiation of ASCs. Promising results were already shown by a study focusing on the differentiation of ASCs on spider silk scaffolds carried out in a bioreactor [18, 19]. Further elastic cartilage in the shape of a human ear was successfully generated in a bioreactor [16].

6 Summary and Conclusion

Summarizing the above, spider silk is a promising material for various biomedical applications. Its *cyto-* and *hemocompatibility* has already been verified while it seems to recommend itself for tissue engineering as biomaterial matrix. Spider silk constructs have also been successfully applied as hernia net, suture material and wound dressing. The next logical step towards common use in the clinical application is consequently the process to approval as a medical product of this unique natural material.

References

1. Vollrath, F.: Strength and structure of spiders' silks. J. Biotechnol. **74**(2), 67–83 (2000)
2. Agnarsson, I., Kuntner, M., Blackledge, T.A.: Bioprospecting finds the toughest biological material: extraordinary silk from a giant riverine orb spider. PLoS ONE **5**(9), e11234 (2010)
3. Risig, A., Nimmervoll, H., Grip, S., Fernandez-Arias, A., Storckenfeldt, E., Knight, D.P., Vollrath, F., Engström, W.: Spider silk proteins—mechanical property and gene sequence. Zool. Sci. **22**(3), 273–281 (2005)
4. Sponner, A., Vater, W., Monajembashi, S., Unger, E., Grosse, F., Weisshart, K.: Compostion and hierarchical organisation of a spider silk. PLoS ONE **10**(e998), 1–8 (2007)
5. Heim, M., Keerl, D., Scheibel, T.: Spider silk: from soluble protein to extraordinary fiber. Ang. Chem. Int. Ed. Engl. **48**, 3584–3596 (2009)
6. Newman, J., Newman, C.: Oh what a tangled web: the medical uses of spider silk. Int. J. Dermatol. **34**, 290–292 (1995)
7. Peers, S.: Golden Spider Silk. V & A Publishing (2012)
8. Work, R.W., Emerson, P.D.: An apparatus for the forcible silking of spiders. J. Arachnol. **10**, 1–10 (1982)
9. Kuhbier, J.W., Allmeling, C., Reimers, K., Hillmer, A., Kasper, C., Menger, B., Brandes, G., Guggenheim, M., Vogt, P.M.: Interactions between spider silk and cells-NIH/3T3 fibroblasts seeded on miniature weaving frames. PLoS ONE **5**(8), e12032 (2010)
10. Schäfer-Nolte, F., Hennecke, K., Reimers, K., Schnabel, R., Allmeling, C., Vogt, P.M., Kuhbier, J.W., Mirastschijski, U.: Biomechanics and biocompatibility of woven spider silk meshes during remodeling in a rodent fascia replacement model. Ann. Surg. **259**(4), 781–792 (2014)
11. Vollrath, F., Barth, P., Basedow, A., Engstrom, W., List, H.: Local tolerance to spider silks and protein polymers in vivo. Vivo Dev. **16**, 229 (2002)
12. Radtke, C., Allmeling, C., Waldmann, K.H., Reimers, K., Guggenheim, M., Brandes, G., Vogt, P.M.: Spider silk constructs enhance axonal regeneration and remyelination in long nerve defects in sheep. PLoS ONE **6**, e16990 (2011)
13. Gellnyck, K., Verdonk, P., Forsynth, R., Almqvist, K.F., van Nimmen, E., Gheysens, T., Mertens, J., van Langenhove, L., Kiekens, P., Verbruggen, G.: Biocompatibility and biodegradability of spider egg sac silk. J. Mater. Sci. Mater. Med. **19**, 2963–2970 (2008)
14. Hennecke, K., Redeker, J., Kuhbier, J.W., Strauss, S., Allmeling, C., Kasper, C., Reimers, K., Vogt, P.M.: Bundles of spider silk, braided into sutures, resist basic cyclic tests: potential use for flexor tendon repair. PLoS ONE **8**(4), e61100 (2013)
15. Kuhbier, J.W., Reimers, K., Kasper, C., Allmeling, C., Hillmer, A., Menger, B., Vogt, P.M., Radtke, C.: First investigation of spider silk as a braided microsurgical suture. J. Biomed. Mater. Res. B Appl. Biomater. **97**(2), 381–387 (2011)

16. Evans, G.R.D., Brandt, K., Widmer, M.S., Lu, L., Meszlenyi, R.K., Gupta, P.K., Mikos, A. G., Hodges, J., Williams, J., Gürlek, A., Nabawi, A., Lohman, R., Patrick, C.W.: In vivo evaluation of poly(L-lactic acid) porous conduits for peripheral nerve regeneration. Biomaterials **20**(11), 1109–1115 (1999)
17. Wendt, H., Hillmer, A., Reimers, K., Kuhbier, J.W., Schäfer-Nolte, F., Allmeling, C., Kasper, C., Vogt, P.M.: Artificial skin—culturing of different skin cell lines for generating an artificial skin substitute on cross-weaved spider silk fibres. PLoS ONE **6**(7), e21833 (2011). https://doi.org/10.1371/journal.pone.0021833
18. Coger, V.: Amélioration d'un tendon artificiel avec de la soie d'araignée. Universität Cergy Pontoise, Frankreich, Masterarbeit (2010)
19. Coger, V., Hillmer, A., Jahn, S., Kasper, C., Kuhbier, J., Schinner, S., Allmeling, C., Reimers, K., Vogt, P.M.: Einfluss von Spinnenseide auf das Tissue engineering von Sehnengewebe. Plastische Chirurgie **10**(Suppl. 1), 98 (2010)
20. Kamara, N., Reimers, K., Aust, M., Guggenheim, M., Vogt, P.M.: Differenzierung von adipozytären Stammzellen (adMSC) zu elastischem Knorpel in Form eines humanen Ohres. Plastische Chirurgie **8**(Suppl. 1), 44 (2008)

The Innovative Power of the Electric Eel (*Electrophorus electricus*)

Jenifer Gifford and Matthew Leming

Abstract Electrogenic fishes have the unique ability to generate electrical discharges from specialized organs composed of electrocytes. These electrical discharges can be used for predatory behavior, inter-/intra- species communication, and electrolocation capabilities. *Electrophorus electricus*, more commonly referred to as the electric eel, is the only large, freshwater, electrogenic fish species belonging to the order Gymnotiformes that is both electroreceptive and capable of generating both weak and strong electrical discharges. Because of this, it has been a source of awe and inspiration for innovators across the globe who have utilized and reverse engineered the biological tools of this organism to develop technologies ranging from the use of the animal as a power source, to generation of electricity from mechanical energy, to the development of artificial, electrogenic cells. Thus, the electrobiology of the electric eel has served as a springboard for applications in industrial, recreational, and healthcare settings and may continue to influence future innovation.

1 Introduction

1.1 Bionics and Innovation

Living organisms have provided a substantial supply of inspiration and creative solutions to many important scientific and engineering based problems. The use of such inspiration has been termed biomimetics or bionics. Many such advances are described in this text and will change the way we not only view the natural world, but also the way we conduct ourselves in a broad range of industries, including healthcare and translational research. In this chapter, we will discuss several innovative technologies that have been developed by talented scientists who looked

J. Gifford · M. Leming (✉)
Children's Hospital of Pittsburgh of UPMC, John G. Rangos Sr. Research Center, 3B11 4401 Penn Avenue, Pittsburgh, PA 15224, USA
e-mail: mtleming@gmail.com

M. Israelowitz et al. (eds.), *Biomimetics and Bionic Applications with Clinical Applications*, Series in BioEngineering,
https://doi.org/10.1007/978-3-319-53214-1_6

to the energy generating properties of electric eel (*Electrophorus electricus*) for inspiration. Because of the potential high impact of the unique physiology present in the electric eel, this chapter seeks to illuminate the potential application of the concepts and processes of its electrobiology that may result in electrogenic cells and wearable power sources.

1.2 Electric Fishes

An electric fish is defined as any fish capable of generating electric fields via a specialized organ composed of electrocytes. These fish are described as electrogenic, which is not to be confused with electroreceptive fish that, while capable of detecting electrofields, do not necessarily generate their own. Electroreceptive fish that are not also considered electrogenic include many species of sharks, skates, rays, and catfish [1]. That being said, most electrogenic fish are also electroreceptive [2]. This chapter will focus on technological advances that inspirationally arose from those that are electrogenic (Fig. 1).

Of more than 30,000 species of fish, only approximately 350 are considered electrogenic and are distributed among the freshwater rivers of Africa and South America and marine environments in the Atlantic [3]. They are divided into two

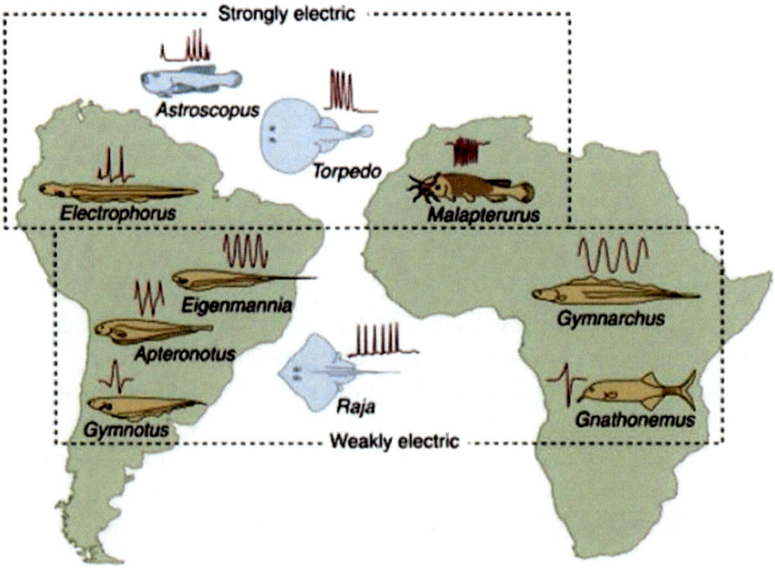

Fig. 1 Electric fish live in the ocean and some live in freshwater rivers of South America and Africa. The figure shows provides a summary of the geographical distribution of electric fishes and shows the location of the electric organ in each fish, and a sample of the waveform of the electric organ discharge (with permission)

categories: those that are weakly electric and those that are strongly electric. Weakly electric fish use their electrogenic capabilities for electrolocation and electrocommunication [2], whereas strongly electric fish, including the electric eel, electric catfishes, and electric rays, can produce electric discharges exceeding hundreds of volts that can be used to stun large prey and ward off predators.

2 Electrogenic Fish and Their Electrobiology

2.1 Weakly Versus Strongly Electric Fish

Weakly electric fish can be characterized further into two groups: variable frequency (buzzers) and constant frequency (hummers). The first is characterized by electrical discharges separated intervals longer than the duration of the discharge with marked increases upon environmental stimulation. The second is characterized by electrical discharges separated by intervals of comparable length to each discharge that are infrequently altered, unless to avoid interference with a frequency similar to their own [4, 5]. They often reside in turbid water and are nocturnal; consequently, they rely on their electric organs for electrolocation to gather information about their environment without the use of vision [3]. Much like the echolocation system used by bats, electroreceptors on these fish can detect alterations in the currents of electrical discharges dispelled by the fish's electrical organs into the surrounding water. This sort of "electrical image" allows the fish to navigate and detect predators and prey. Electric organs of weakly electric fish are also used for communication, which can be done inter- or intra- specially, as discharge pattern and pulse are generally species specific and can transmit information, such as the sex of another nearby electric fish [4] (Fig. 2).

Functionally, the organs of a strongly electric fish are capable of generating electrical discharges of substantial output used for stunning large prey or resisting and warding off potential predators. Because of the output, strongly electric organs dispel charge in isolated bursts. In some cases, such as that of the electric eel, strongly electric fishes possess both strongly and weakly electric organs for both electrogenic and electroreceptive purposes [4].

As both strongly and weakly electric fishes have harnessed the capacity of electrocytes to generate electrical charge for a variety of means, scientific ingenuity may also provide for the application of electrocyte-inspired technology to a variety of innovations.

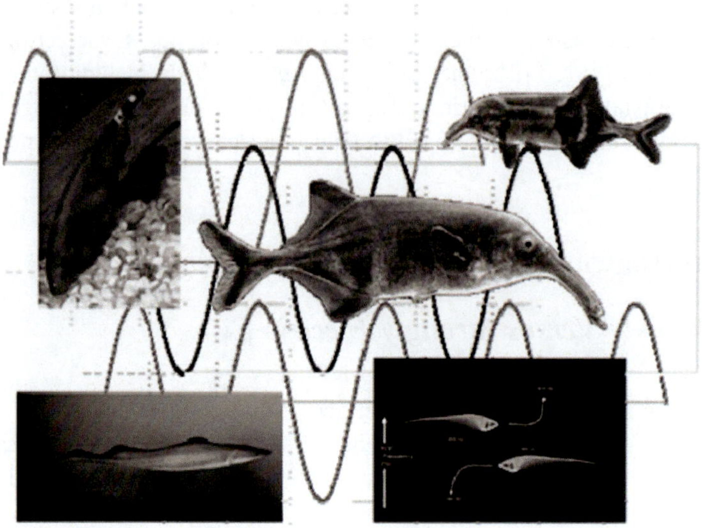

Fig. 2 Different weakly electric fishes (with permission)

2.2 *Electrophorus Electricus*

Electrophorus electricus, or the electric eel, is the largest freshwater species of the order Gymnotiformes, is both electroreceptive and electrogenic, and is capable of generating both low and high voltage discharges through three distinct pairs of electric organs: the main electric organ, the Hunter's organ, and the Sach's organ [6, 7]. The main organ and the Hunter's organ are responsible for strong electric organ discharges [7] and are used for predatory behavior, such as immobilizing their prey or revealing hidden prey by using their electrical discharge to affect the motor neurons of nearby prey, causing whole body contractions [8]. The Sach's organ emits low electric organ discharges [7] and is more associated with an electrolocation function, as these fish are also nocturnal and occupy murky environments, as previously described (Fig. 3).

While there are a number of fish species that are capable of generating electric discharges, the electric eel is the largest and only freshwater fish species with the intense capacity to generate both low and significantly high electrical discharges (as high as 600 V) from acquired energy. These qualities make it an ideal model for study by scientists and innovators striving to harness new, efficient strategies for energy generation [9].

Fig. 3 Electric eel as an elongated, cylindrical body, typically growing to about 2 m in length, and 20 kg. Their coloration is dark gray-brown on the back. They have no scales. The mouth is square, and positioned at the end of the snout

2.3 A Simplified Overview of Electrocyte Physiology

It is the unique biology and physiology of the electrogenic organs in the electric eel that have made it a model of significance in the field of biomimetics, as these organisms are capable of converting the energy they acquire into useable electrical currents (Fig. 4).

The electric organ of the eel is composed of stacks, or columns, of specialized plate-like muscle cells termed electrocytes. Each stack may contain thousands of these electrogenic cells, which are stacked resembling a voltaic pile in proximity to a motor neuron at the end of each stack. In a position of rest, sodium/potassium pumps within the cells export positively charged sodium and potassium ions, providing an overall negative charge within each electrocyte. Upon stimulation, acetylcholine signals from the motor neuron open the voltage gated sodium/potassium channels, allowing positive ions to flow back into one end of the cell. This provides for a positively to negatively charged gradient within the cell, through which the current flows and is ultimately dispelled from the animal [4, 7].

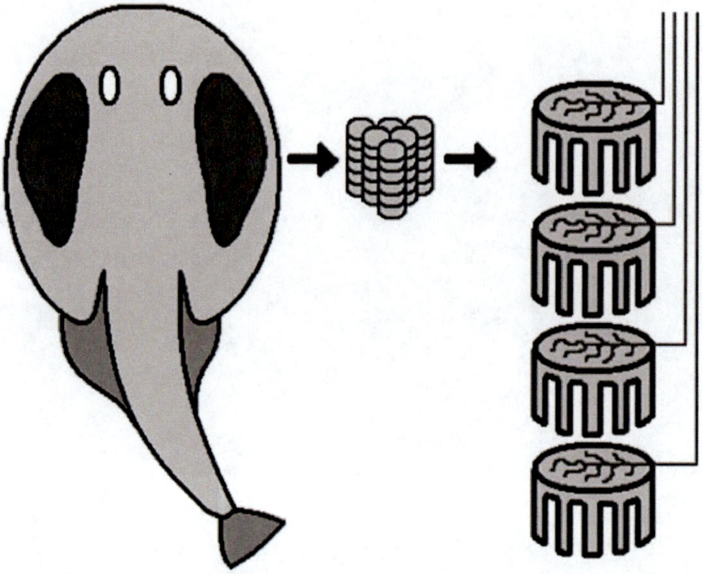

Fig. 4 The weakly electric fish generates Electric Organ Discharge (EOD) with specialized compartments called electric organ. Almost all the weakly electric fishes have electric organs derived from muscle cells (myogenic), there is one exception the Apteronotidae, the electric organs derived from nerve cells (neurogenic) (with permission)

3 Innovations

3.1 Electrogenic Skin

There are a broad range of applications that can be imagined for both wearable and non-wearable electrogenic "skin," with future applications relating to non-wearable technology likely far surpassing the current limits of scientific innovation. These include applications to robotics, healthcare, and wearable devices used in both recreational and healthcare settings. Wearable technologies have recently become more popular recreationally and have flourished as a useful tool in healthcare, when used as body sensors for measurement of vital signs and head mounted display devices for assistance during surgical procedures [10].

In 2014, a new paradigm of energy harvesting was developed using triboelectric nanogenerators (TENGs). These TENGs functioned by directly converting mechanical energy (i.e. motion, vibration, mechanical stimuli, physical touching, and biological movement) into electrical energy [11]. Later, in 2016, the invention of flexible capacitors with high output voltages allowed for the development of flexible electrogeneration [12]. Through the use of these two advancements, and in mimicking the skin of the electric eel, investigators have created a flexible skin-like material that generates electricity through mechanical manipulations. These devices have the ability to convert mechanical energy into electricity [13]. This group has contributed a "mechanically durable and resilient nanogenerator by using triboelectric effects; skin-like triboelectric nanogenerator (SLTENG)" [13].

There are many potential uses for a synthetic second-"skin" that can generate consistent, renewable electricity. Currently these devices are being investigated for their potential use in robotics, wearable/implantable healthcare devices, etc. However, future applications may be very diverse and span many industries.

3.2 Electric Cells for Biomedical Devices/Implants

In 2008, there was an interest in creating artificial electrocytes for use in small scale devices (e.g. implantable medical devices). Xu and Lavan described how artificial electrocytes could generate electricity by mimicking the ion channel arrangement of electrocytes [14]. In principle, artificial electrocytes would take advantage of the innate nanoscale conductors that manifest as ion channels and pumps in cellular membranes. They further observed that optimization of this method of artificial electrocyte generation had the potential to produce more energy more efficiently for these types of small scale devices. However, further development of this technology is necessary to build it into a format that is commercializable.

4 The Future of Electric Fish

4.1 Diverse Applications for Energy Sources

The future of applying the electrogenic properties of electric fish to sources of energy generation has the potential to make significantly positive impacts. As the electric fishes themselves make use of their energy generation for a multitude of tasks, future innovation, through the mimicking the energy production of the electric eel, can lead to the development of novel inventions related to large scale energy generation (e.g. lights, autos) or devices used in combination with existing power generating techniques. An example of this could be the production of in vitro and in vivo "biobatteries." Through exploration of the electric eel genome, it may be possible to genetically engineer single celled photosynthetic eukaryotes, such as algae or yeasts, to convert sunlight into usable electric currents. Furthermore, this concept could be applied to the conversion of non-electrogenic cells in vivo to electrogenic cells, in order to power implanted medical devices [9].

4.2 *Inspiring the Next Generation of Innovators*

Beyond the aforementioned technological advances, other applications exist that are not readily commercializable, yet they have been used to educate and captivate broad audiences and the next generation of potential innovators. At the Tennessee Aquarium in Chattanooga lives a particular electric eel known to aquarium visitors as Miguel Wattson. Miguel is housed in a manner that allows the electricity he generates to power pre-written posts to his Twitter feed. Other individuals have taken advantage of the electrogenic nature of these animals in very creative ways. Most notably, at an aquarium in Kamakura, Japan, an electric eel is used to power the lights on a 6 ft Christmas tree during the Christmas season. The light hearted nature of these applications demonstrates that the potential for application is literally only limited by the imagination.

5 Conclusions

Electrogenic fishes, particularly the electric eel, serve as an inspirational biological model for bionic innovation in the realm of energy generation. Whether through the use of wearable devices capable of generating electricity from mechanical energy or genetically engineering a cell to acquire electrogenic properties, exploring and harnessing the biological processes that govern electrogenicity will impact the future of energy generation.

References

1. Moller, P., Electric fishes: history and behavior. Chapman and Hall, London (1995)
2. Alves-Gomes, J.A.: The evolution of electroreception and bioelectrogenesis in teleost fish: a phylogenic perspective. J. Fish Biol. **58**, 1489–1511 (2001)
3. Nelson, M.E.: Electric fish. Curr. Biol. **21**(14), R528–R529 (2011)
4. Bennett, M.V.L.: Comparative physiology: electric organs. Annu Rev. Physiology **1970**, 471–528 (1970)
5. Watanabe, A., Takeda, K.: The change of discharge frequency by A.C. Stimulus in a weak electric fish. J. Exp. Biol. **40**, 57–66 (1963)
6. Mermelstein, C.D.S., Costa, M.L., Neto, V.M.: The cytoskeleton of the electric tissue of *Electrophorus electricus*. Anais da Academia Brasileira de Ciencias. **72**(3), 341–351 (2000)
7. Ching, B., Woo, J.M., Hiong, K.C., Boo, M.V., Choo, C.Y.L., Wong, W.P., Chew, S.F., Ip, Y.K.: Na +/K + -ATPase α-subunit (nkaα) isoforms and their mRNA expression levels, overall Nkaα protein abundance, and kinetic properties of Nka in the skeletal muscle and three electric organs of the electric eel, Electrophorus electricus. PLoS One **10**(3) (2015)
8. Catania, K.: The shocking predatory strike of the electric eel. Science **346**(6214), 1231–1234 (2014)

9. Albert, J.S., Zakon, H.H., Stoddard, P.K., Unguez, G.A., Holbert-Albert, S.K.S., Sussman, M. R.: The case for sequencing the genome of the electric eel Electrophorus electricus. J. Fish Biol. **72**, 331–354 (2008)
10. Igmal, M.H., Aydin, A., Brunckhorst, O., Dasgupta, P., Ahmed, K.: A review of wearable technology in medicine. J. R. Soc. Med. **109**(10), 372–380 (2016)
11. Wang, Z.: Triboelectric nanogenerators as new energy technology and self-powered sensors—Principles, problems and perspectives. The Royal Society of Chemistry. Faraday Discussions (2014)
12. Sun, H., Fu, X., Xie, S., Jiang, Y., Peng, H.: Electrochemical capacitors with high output voltages that mimic electric eels. Adv. Mater. **28**, 2070–2076 (2016)
13. Lai, Y., Deng, J., Niu, S., Peng, W., Wu, C., Liu, R., Wen, Z., Wang, Z.: Electric eel-skin-inspired mechanically durable and super-stretchable nanogenerator for deformable power source and fully autonomous conformable electronic-skin applications. Adv. Mater. (2016). https://doi.org/10.1002/adma.201603527
14. Xu, J., Lavan, D.: Designing artificial cells to harness the biological ion concentration gradient. Nat. Nanotechnol. **3**(11), 666–670 (2008)

Morphological Study of the Infrared Sensory Pits of Pit Viper, Python and Boa Snakes

Birgit Weyand, Meir Israelowitz, Matthias Reuter, Sabine Bohlmann, Robert Wagner, Syed W. H. Rizvi, Chistoph Gille, Kerstin Reimers, Peter Vogt, and Herbert P. von Schroeder

Abstract Different families of snakes possess specialized infrared sensory organs which have evolved for thermoregulation as well as prey and predator detection. The sensor, a specialized "pore" found in the keratinized epidermal layer, is activated by infrared thermal radiation and transmits signals via the neuronal system through the trigeminal nerve. Here we present findings from anatomical and physiological studies of the infrared sensor of pit viper, python and boa using histology, optical coherence tomography, laser microdissection and capillary electrophoresis to identify and characterize a lipoprotein and its important role during sensor activation. Furthermore, the mechanism for signal detection and enhancement as well as reduction of background noise was simulated using neural network analysis based on the findings from histology. A high density of free nerve endings and multiple neural (i.e. dendrite) connections in the sensor organ are crucial for signal detection and processing and work as a filter to overcome background noise.

Keywords Morphology · Pits · Infrared sensors · Histology · Neural networks · Distribution systems

---

B. Weyand (✉) · K. Reimers · P. Vogt
Department of Plastic, Hand and Reconstructive Surgery, Hannover Medical School, Hannover, Germany
e-mail: birgit.weyand@mh-hannover.de

M. Israelowitz · S. W. H. Rizvi · C. Gille · H. P. von Schroeder
Biomimetics Technologies Inc, Toronto, ON, Canada

M. Reuter · S. Bohlmann
Technical University, Clausthal-Zellerfeld, Germany

R. Wagner
Medical School University of Pittsburgh, Pittsburgh, PA, USA

H. P. von Schroeder
University of Toronto, Toronto Western Hospital, Toronto, ON, Canada

M. Israelowitz et al. (eds.), *Biomimetics and Bionic Applications with Clinical Applications*, Series in BioEngineering,
https://doi.org/10.1007/978-3-319-53214-1_7

1 Introduction

Visible light (400–700 nm) is a narrow spectrum of electromagnetic radiation in which we and other creatures sense their environment [1]. Evolution of other animals has exploited sensory abilities in other parts of the spectrum. For example, infrared wavelengths are sensed by specialized receptors in the fire chaser beetle *Melanophila acuminata* [2–5], and in three subfamilies of snakes: pythons (family Pythonidae), pit vipers and specific boas (boidae) [6]. The natural infrared sensors of these snakes have the ability to detect much less than one degree differences in temperature [7] and utilize it to detect their prey. Research over the past 30 years on these species has investigated the physiological [8] and physical models [9] of infrared sensory organs in snakes and has found them to have similar function, but differing morphological features.

In all snakes, the infrared pits are innervated by the trigeminal nerve [10, 11] which transmits the neuronal signal [12, 13]. In pythons and boas the nerve endings are at the base of pits, but pit vipers have the nerve endings within a membrane in the middle of the pits [14].

In pythons the *thermoreceptive* labial pits are depressions that are located around the mouth, mostly on the upper lip. These pits have orifices with an approximate size of 1–2 μm [7] and are optically described as a pinhole model [5] but lack the high resolution that a camera may have. The size of the porous orifice, where the sensor is located, limits the length of the electromagnetic wave being detected. Amemiya et al. [15] and Campbell et al. [16] studied the structure and function of the infrared pits in Burmese pythons using scanning electron microscopy. They described pores that selected only long wavelengths and reflected the short waves [15, 16]. By taking advantage of the sizes, the snakes are able to filter the specific wave length and improve the efficiency of the work related the porous pits located in the keratinized epidermal layer [17–19].

In pythons the base of the pits are lined with a membrane that is vascularized for cooling purposes and contains the trigeminal nerve endings that directly function as infrared detectors [20–22]. The intensity of the infrared radiation in the python depends on the geometric shape of the supralabial pits and the caudal pits [6, 18, 22–24]. The difference in the intensity and the response to the detection of the heat correlates with those structures [14, 25]. The correlation with the response of the neurons depends on the geometric shape given by the target [26].

The pit vipers are in-part defined and classified by their infrared sensing pit on each side of their head between the eye and the nostril in the loreal region. Unlike boas and pythons, the vipers' pits contain a membrane that divides each pit into two chambers. It is this membrane that contains the nerve endings from the trigeminal nerve; the region is also rich in mitochondria within the neural cells. The two chambers are connected by a narrow duct that is opened or closed by surrounding muscles to control the air pressure [7].

The heat sensitivity in the snake employs an ambush or *sit and wait* strategy. In the process of capturing the prey, the snake has to follow a set of movements [27].

The first movement is for orientation and the second movement is to correct the striking distance. The strike time is 35 ms [7].

During the orientation phase the boa turns its head towards the prey (e.g. a mouse) and, depending entirely on the heat of the prey, automatically calculates the strike zone. In contrast, the sensor of the python has a lower sensitivity for heat and is more reliant on olfactory signals. In 1999, Chiba et al. described a high concentration of mitochondria in the neurons of the pit sensor which represents a high neural activity [28]. The neurites also have a high concentration of calcium binding protein for optimal function of heat-sensitive ion channels at temperatures between 25 and 30 °C [21].

The aim of our study was to bridge the gap between morphological analysis of the pit sensor organ and its functional requirements. Using neural network simulation, we were able to simulate the activity of the multi-layer dendrite composition in order to overcome the thermal noise and in order to localize the switch which translates the sensor signal into an active response.

2 Materials and Methods

2.1 Snakes

Tissue samples were obtained from two Crotalinae pit vipers and one true python. The snakes were obtained with assistance from the Pittsburgh Herpetology Society and through private breeders. The snakes where kept at room temperature and under strict conditions, and were fed once a week, using a protocol from Dr. Wagner, veterinarian.

2.2 Scanning Electron Microscopy (SEM)

The snake sensory pits were prepared in 70 and 100% ethanol and processed for freeze-drying. The pits were assessed using a Philips XL 30 FEG SEM scanning electron microscope. Micrographs where taken with a Philip XL 30 FEG SEM and samples were prepared in 70%/100% ethanol and processed for freeze-drying.

2.3 Optical Coherence Tomography (OCT)

Sensory pits were analyzed in situ via optical coherence tomography (OCT) developed in-house at Carnegie Mellon University and controlled by a software (computer) [29]. Each sample was exposed to a low coherence *superluminescent* laser with a wave length of 1300 ± 15 nm and with an output power of 300 μW. The light from the interferometer was transmitted by a single mode optical fiber and the depth scanning was produced by piezoelectric modulation of the fiber length. An image was obtained by scanning both in a lateral direction (X axis) and deep into the sensory pit (Z-axis) by the interferometer. The resultant images acquired a resolution of 450 × 450 pixels.

2.4 Histology

Frozen sections of the sensory pits were processed by slicing 5.00 μm thin sections, and stained using Azure methylene blue eosin [30] dissolved in phosphate buffer solution. Oil Red O [31], Sudan I [32], Sudan II [29], Sudan III [33] and Black Sudan [34, 35] were used for staining lipids and phospholipids.

2.5 Confocal Microscopy

In vitro assessment of the snakes' sensory pits was undertaken with a Nikon confocal microscope. For excitation, a multi-argon laser (40 milliwatts; wavelengths: 458, 488, 514 nm) and a helium-neon (red) laser (10 milliwatts; wavelength 633 nm) were used. Average, low-pass and high-pass filters were utilized and a cut-off frequency slider was used to vary the range of the frequencies that were passed by the filter. A monochrome CCD camera (Hamamatsu Digital Camera CA 742-95) was employed for imaging [36].

2.6 Laser Microdissection

The snake sensory organs were dissected and embedded in the OCT compound and frozen at −80 °C for cryostat sectioning followed by 2% Giemsa nucleic acid stain at the Beckman Institute (CA, USA). Sections were desiccated under vacuum for 10 min, and placed under a stereoscope [37]. A drop of water was applied to remove the OCT mounting compound. Sensory organs were then microdissected using Beckman Laser Institute laser scissor/tweezers system and the optical characteristics of the dissected components were determined. Microscopy was performed on dissected sections [38–42].

2.7 Capillary Electrophoresis

Capillary Electrophoresis (CE) was used for the separation and subsequent identification of the protein components from the LCM extracted tissue samples. CE is an alternative and novel format for both liquid chromatography and electrophoresis. The molecular weight range of analyses separable by CE is enormous, ranging from 3 to 100,000,000 Daltons (1). Detection was performed with the assistance of covalently linked fluorescent dyes, which improve detectability to the *attomole* range [43]. CE test runs employing Cy5-BSA in concentrations close to 60 *femtograms/microliter* have been positively identified in the CE chromatogram. Further analysis of the fluorescent-labelled BSA via MALDI/TOF mass spectrometry showed no interference from part of the dye with respect to ionization or detectability of the protein. In comparison to control experiments, the stark difference was a molecular weight shift in the Cy5-BSA dependent on the D/P (Dye/ Protein) ratio [44]. This allows for interfacing of CE (LC = Liquid Chromatography) to the mass spectrometer via the electrospray, increasing therefore the sensitivity of detection and precision of analysis with the added benefits of ease for automation, precision and ruggedness offered by CE [45] (Fig. 8).

2.8 Software

The image analysis of the nerve distribution was done in-house using a neural network system which had previously been developed for radar visualization. The software analysed the morphological structure and provided information about the structure. The neural network began without any preconceived assumptions of the system and learned from the image recognition to highlight structures [46–54].

3 Results

Pit vipers possess one large pit organ on either side of the head. Figure 1 shows an X-ray of a viper focusing on the loreal area between the eyes and nostrils where the pit organ is located. In comparison, boas and pythons have three or more comparatively smaller labial pits lining the upper and sometimes the lower lip, in or between the scales. Generally, vipers have a whole size of 30–200 cm whereas pythons and boas can reach whole sizes of 1–10 and 2–11 m, respectively.

Figure 2 shows the jaw of a python with the pit organs at the jaw. All snakes have specialised locations where these receptors are found [55].

Figure 3 shows the OCT image at the location of the python sensor. The penetration of the OCT was 7.9 mm by 4.3 mm.

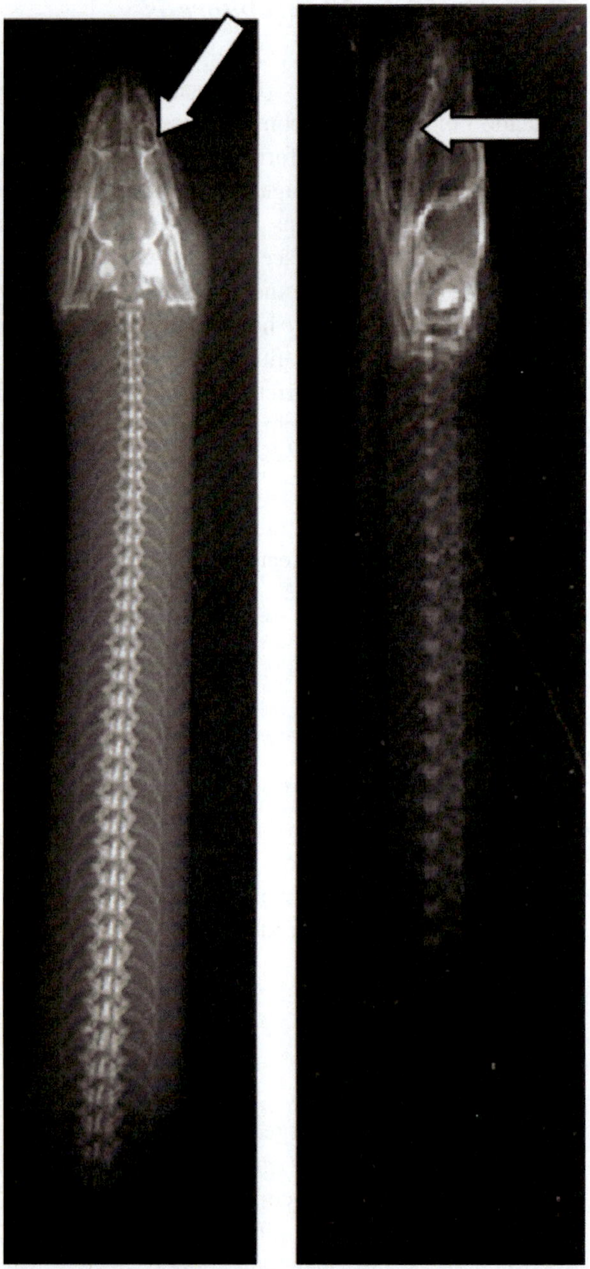

Fig. 1 X-ray of the viper before sectioning with focus on the vipers head where the loreal pit sensor is located between the eye and the nostrils and positioned deep into the maxillary cavity. (X-ray graciously provided by Dr. Robert Wagner, V.M.D. University of Pittsburgh)

Results of the histological examination are shown in Fig. 4: Fig. 4a shows a cross-section of the viper sensor region, where a deeply infolded epithelial lined region is surrounded by nerve endings (arrowheads). Adjacent a region is found, purple stained by Oil Red O, indicating a lipid-rich region such as a dendrite rich layer, which is shown in Fig. 4b in more detail. Oil Red O stain is a lysochrome

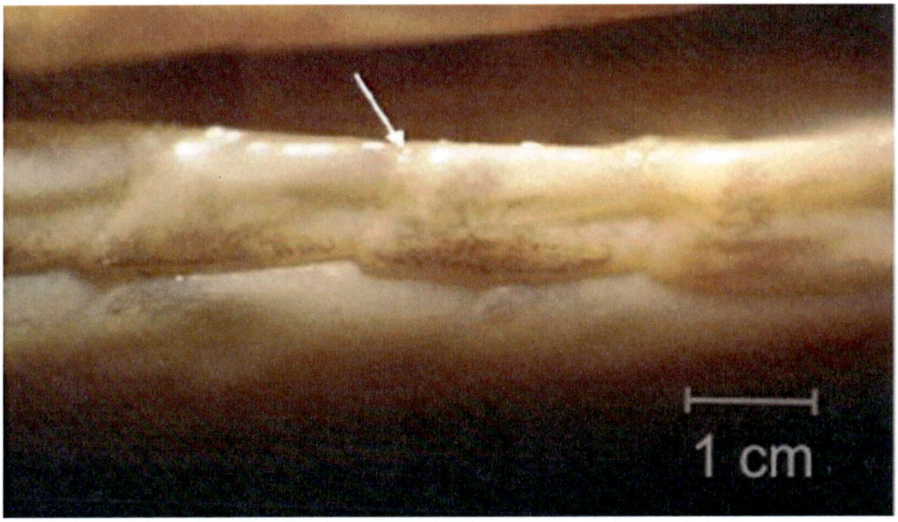

Fig. 2 Bright-field RGB image of a python jaw with the arrow pointing to the pit receptor location in the labia—in this case lower labia

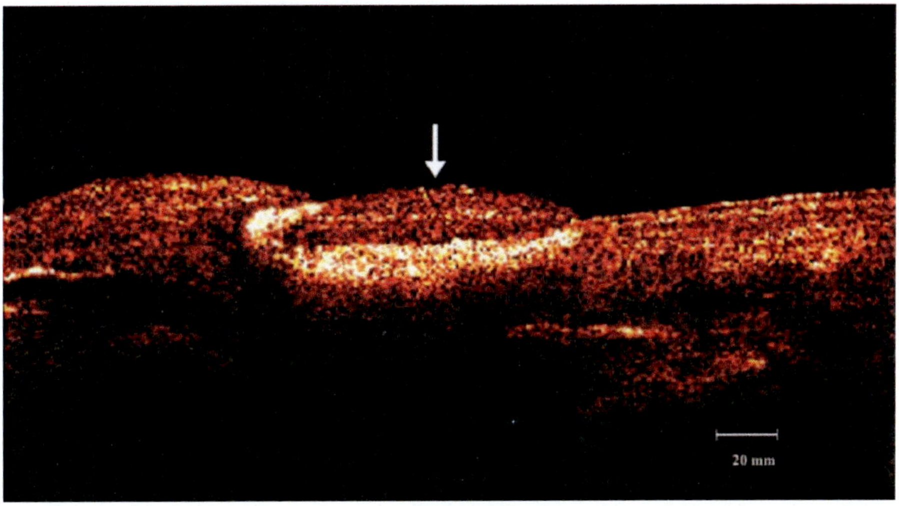

Fig. 3 OCT Image of the python pit in situ. Image size 7.9 mm × 4.3 mm. The white arrow identifies the porus location

diazo dye used for lipids in frozen sections or paraffin sections. It has the appearance of a red powder with maximum absorption at 518 nm.

Sections of the python sensor region are shown in Fig. 5a, b, both with staining of azure-eosin methylene blue. The sensor region as shown in Fig. 5a is lined by an inner "dotted" epithelial layer, where multiple purple-stained open nerve endings are found close to capillaries and to the epithelial surface layer (arrows). Figure 5b

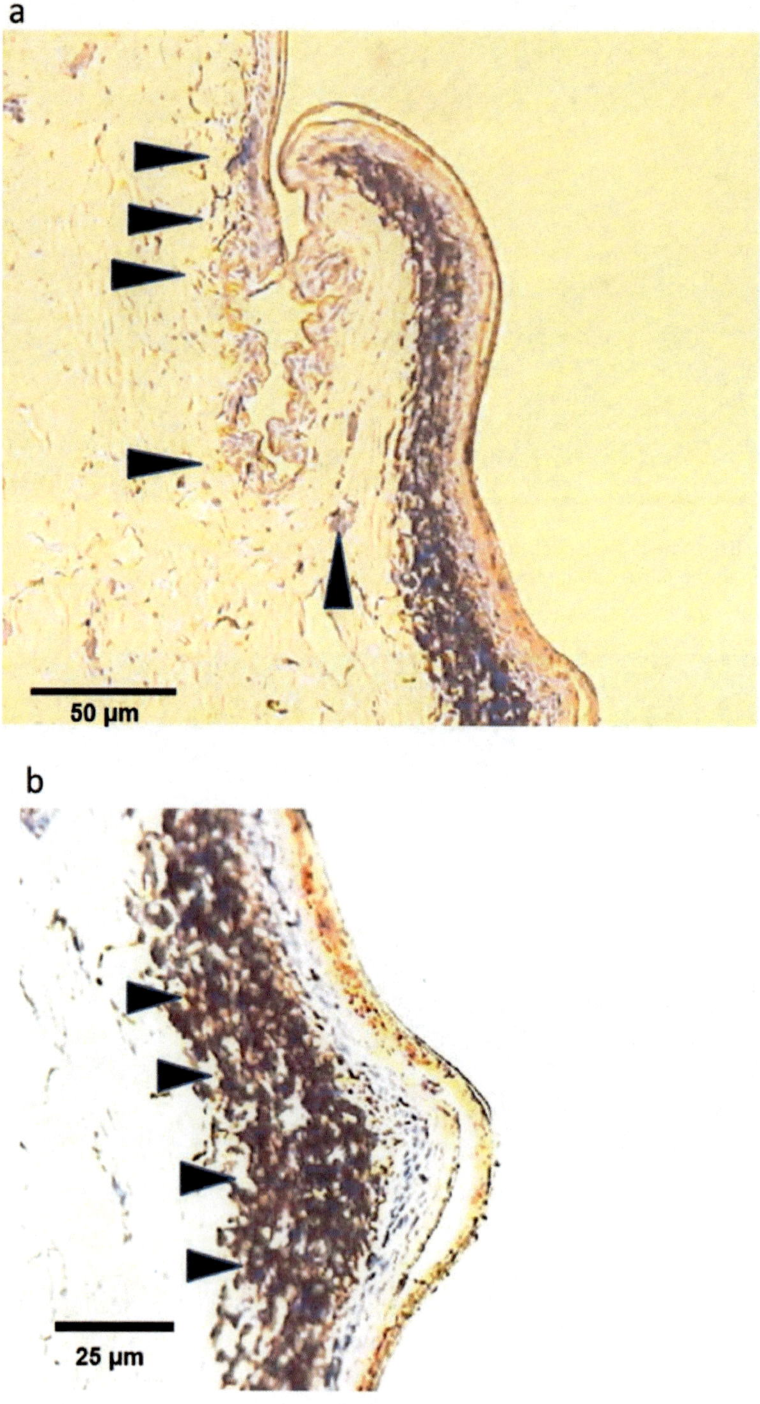

Fig. 4 **a** Cross sections of viper sensor region stained with Oil-Red-O demonstrates an infolded epithelial lined region surrounded by nerve endings and a lipid-rich multilayer region underneath the epithel. **b** A lipid-rich multilayered region of the viper sensor is stained dark purple by Oil-Red-O staining

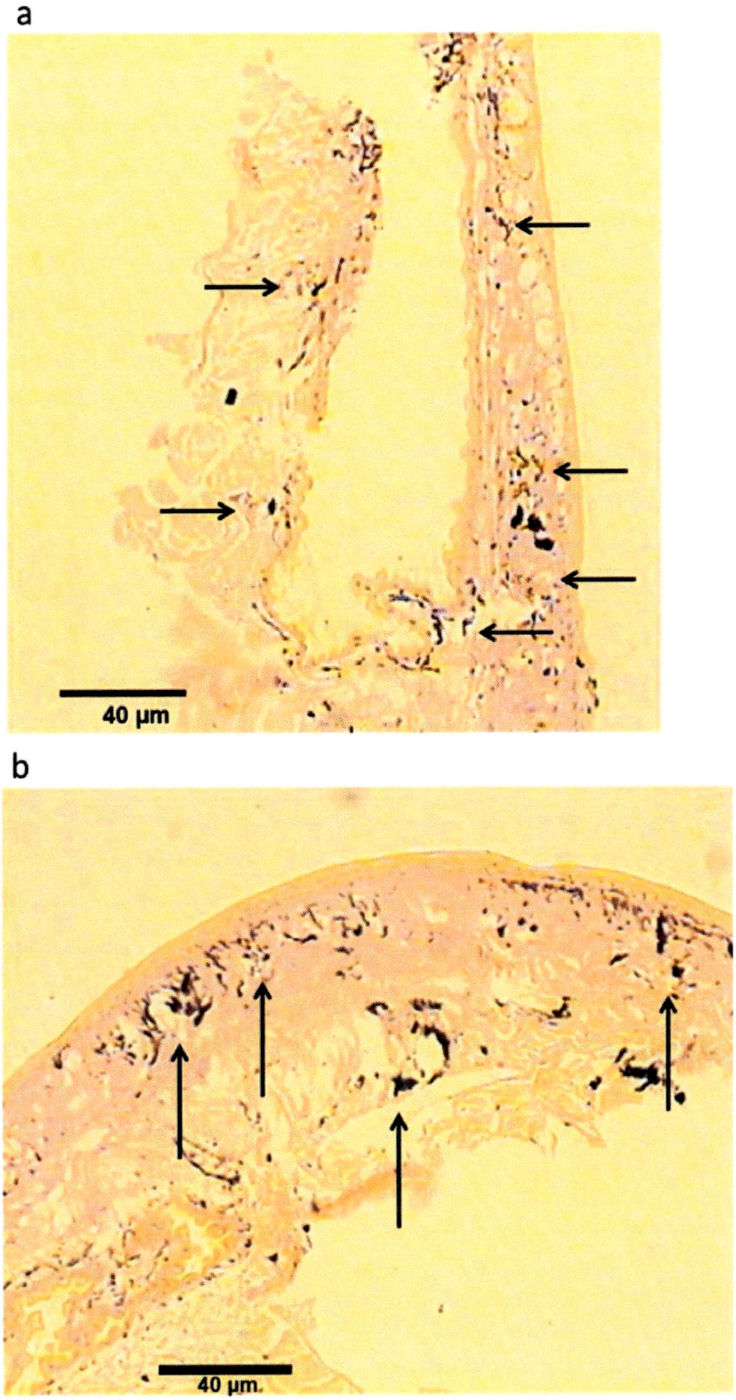

Fig. 5 **a** and **b** Cross-sections of the python sensor region stained with azure-eosin methylene blue. Purple stained nerve endings (arrows) are found close to the epithelial layer and close to capillaries within the connective tissue layer

shows another part of the python sensor region, where purple stained nerve endings are located in proximity to the epithelial layer and also to capillaries in the connective tissue layer (arrows).

Figure 6 shows cross-sections of the python sensor region stained with Sudan staining for lipid detection. Sudan dyes have a high affinity to triglycerides, lipids, and lipoproteins. In Fig. 6a, there are regions underneath the epithelial layer stained with yellow-brownish pigment which represents a layer of dendrites. More centrally within the connective tissue, there is a structure resembling a nerve structure surrounded by a myelin sheet, which are most likely belonging to fibres of the trigeminal nerve. Figure 6b shows a closer look at the lipid-rich layer of nerval dendrite structures which lay adjacent to the epithelial layer. The multiple layers of dendrites serve not just as a detector, but minimize thermal noise encountering with signal processing for identification of a signal against a background which in the case of this sensor system is the environment [56].

Figure 7 demonstrates sections of the boa sensor region which were prepared using Laser Capture Microscope. Figure 7a, b show a cross section of the sensor region of the boa. With Laser Capture Microscopy, the laser cuts were prepared with a laser spot of 7,5 μm with a pulse power of 75 μW, pulse width of 1,5 μs and a threshold voltage of 235 μV total pulses with an estimated transfer per pulse at 90% and the estimated volume of tissue dissected at $7.7 \times 10^{-8}\ \mu^3$ [57]. The laser spot of 7.5 μm is shown in Fig. 7c. The structural composition shows similarities to the sensor regions of viper and python with a dense lipid-rich layer underneath an epithelial outer border layer. Snake species like the boa initially use olfactory sensors, whereas vipers and pythons only use their pits for detection.

Figure 8 shows the results of Capillary Eletrophoresis used for separation and subsequent identification of the protein components from the Laser Capture Microscopy extracted tissue samples from the boa sensor region. The resulting peak at 5.21 min matches a lipoprotein (Fig. 8).

4 Discussion

The near infrared snake detector structure uses a mechanism of detection that is based on lipoproteins and a dendritic neurological system for exposure and signal transduction. But regardless of the detection mechanism, the detector must overcome the background noise [58, 59]. The noise by definition is the background versus the real signal, and the noise is independent of how good a detector is [60–63]. While the infrared sensor of snakes lies in the range of the near infrared spectrum, there are other examples in nature such as the fire chaser beetle *M. acuminata*, which sensor is working within the middle infrared spectrum [5].

In the case of *M. acuminata*, noise is overcome via multiple sensors [44, 45, 58–63]. But in the snake, detection works via a multiple layers of dendrites in a porous region [22].

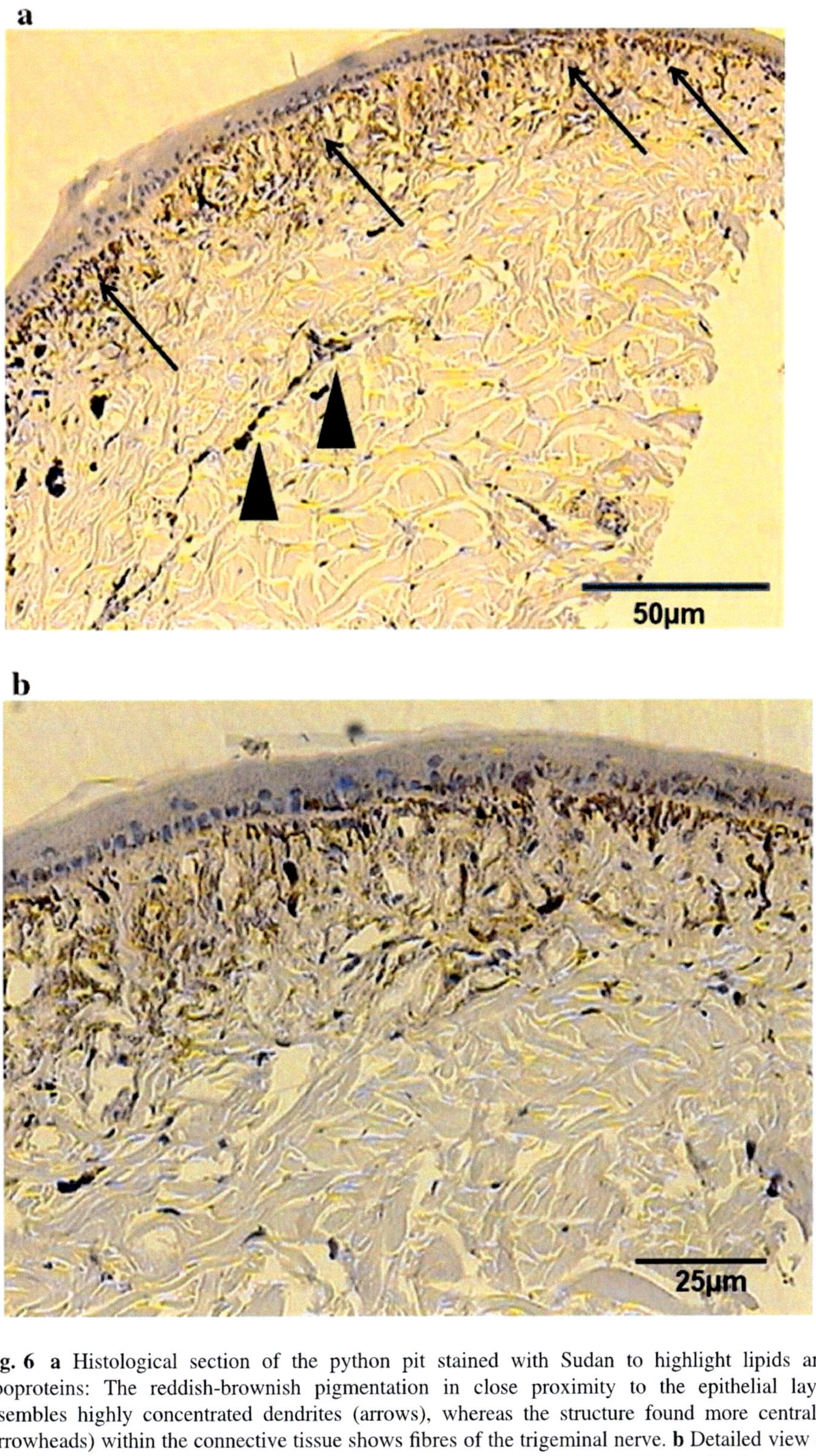

Fig. 6 **a** Histological section of the python pit stained with Sudan to highlight lipids and lipoproteins: The reddish-brownish pigmentation in close proximity to the epithelial layer resembles highly concentrated dendrites (arrows), whereas the structure found more centrally (arrowheads) within the connective tissue shows fibres of the trigeminal nerve. **b** Detailed view of the lipid-rich dendrite layer (arrows) lying adjacent to the epithelial layer (Sudan Black stain)

a

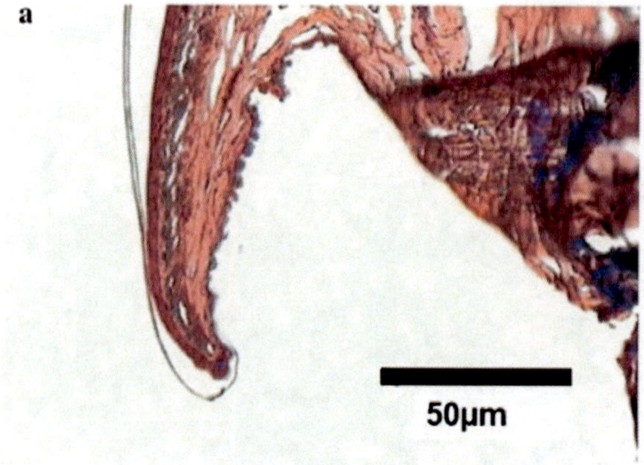

b

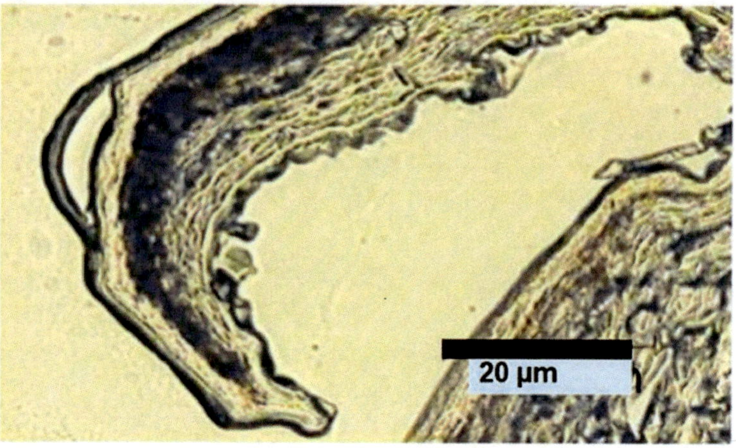

c

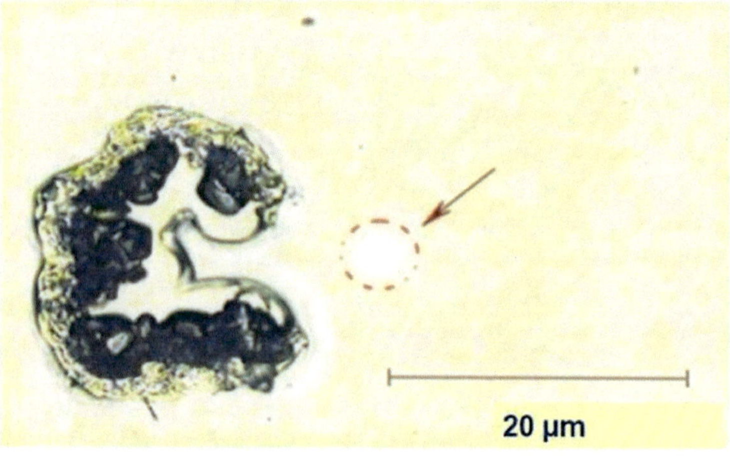

Fig. 7 **a** Histological section of the boa sensory region used for the laser cuts at × 60. **b**. Prepared section in ethanol before was Laser Capture Microscope (LCM). **c** Brightfield image showing cross section of Boa pit prior to resection using Laser Capture Microscope (LCM), Spot size = 7.5 microns

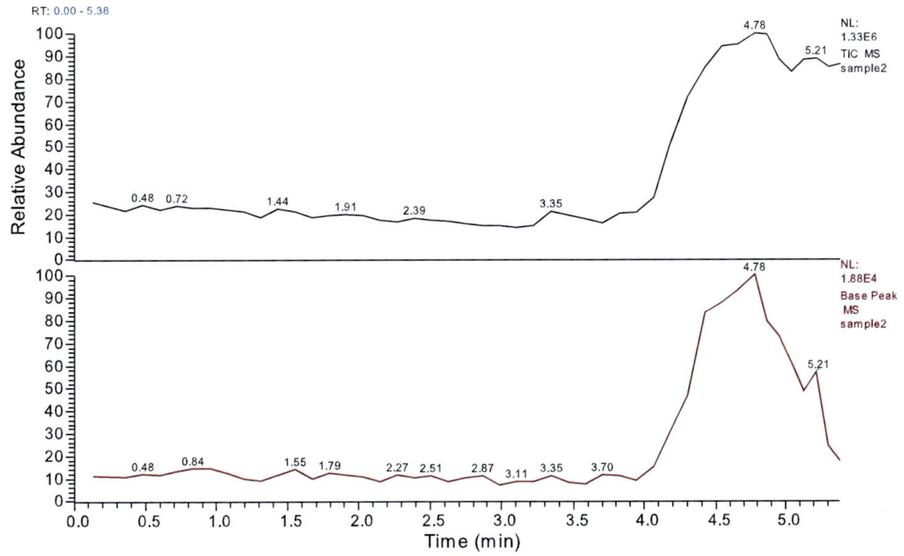

Fig. 8 CE run and base sample with both lines showing peak at 5.21 min representing lipoprotein

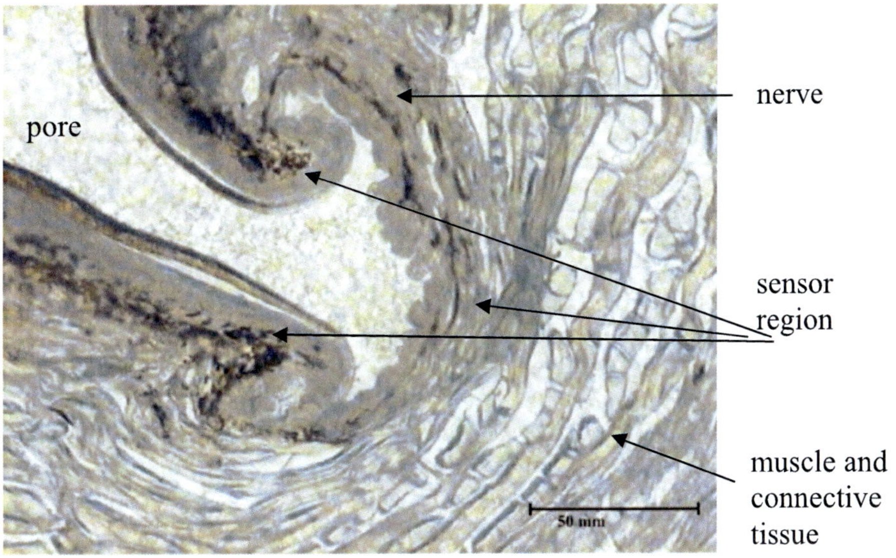

Fig. 9 Histological sections showing the structure of the sensor of a viper pore. The darker colour shows the exposed nerve, that senses the infrared wavelength

By using artificial neural network analysis applied to an initial image, the analysing software creates a multi-layer for each pixel of the image, and considers darker spots as dendrite activity. Based on the dendrites, a perceptron layer is being built, where multiple layers show not only the activity but overcome any noise by averaging the signal which is being spread by multiple layers of dendrites [64, 65]. Histological analysis has shown a multi-layered dendrite layer within the sensor region of pythons, boas and vipers, as seen above.

Figure 9 shows a histological section of the sensor region of the viper pore, which serves as basis for results of the neural network analysis shown in Fig. 11. Figure 10 shows the simulation results of the neural network analysis based on histological sections of the python sensor region shown in Fig. 6b.

The neural network analysis highlights density concentration of the connectivity between the dendrites where white spaces highlight higher concentration of connectivity.

The neural network highlights the high vascularization in the area of the sensor which corresponds to the high connectivity of dendrites found in the sensor region, representing high activity in the specific sensor area [66].

The neural network not only defines the biological structures, but gives a three-dimensional structural overview and provides information regarding the biological functionality [67, 68]. The concentration of the connective networks of the dendrites shows the complexity of the sensory system, and is required to overcome the thermal noise [56].

The neural network highlights structural contrasts as seen in Fig. 10d and specific differences in different intensities are identified: Fig. 10e–h shows examples of how the neuronal network by using the same contrasts can highlight different aspects of the sensor by which helps to identify nerve fibres, dendrites as well as the interconnectivity of the sensor structure.

Figure 9 gives an overview of the porous sensor region of the viper. Based on this histological section, the neural network can analyse an overview of the sensor structure, which is demonstrated in Fig. 11: The location of the lipoprotein is here shown in yellow, the dendrites are shown in red, where the interaction between the infrared radiation received and the switch to convey the information is taking place.

More details of the neural network base analysis are shown in Fig. 12: The contrast structure of the neural network base analysis is shown in Fig. 12a. Figure 12b shows a three-dimensional structural analysis where the nerve endings in the bottom are more pronounced, whereas the results of the neural network in Fig. 12c highlights lipoprotein and dendrite activity in red in contrast to the deeper nerve endings in purple.

Fig. 10 **a** The neural network analysis highlights the connective section between the dendrites. ▶ **b** The areas of high connectivity with saturation of the dendrites in the sensor area. **c** The vascular concentration is highlighted and the concentration correlates to the regions of high connectivity of the dendrite region. **d** The distribution of the network is highlighted as the bright spaces, and the darker spaces are where dendrites are located. **e** The contrast is the basis for the neural network contact analysis. **f** The result of the neural network analysis highlights the dendrites in red and shows the extensions of the dendrites in yellow and the nerves in the section of the sensor in purple. **g** The neural network concentrates in the location of the dendrites and at the connections between their tips and the nerves. **h** The neural network highlights all structures by showing the ridges and connectivity between the nerves and dendrites

a

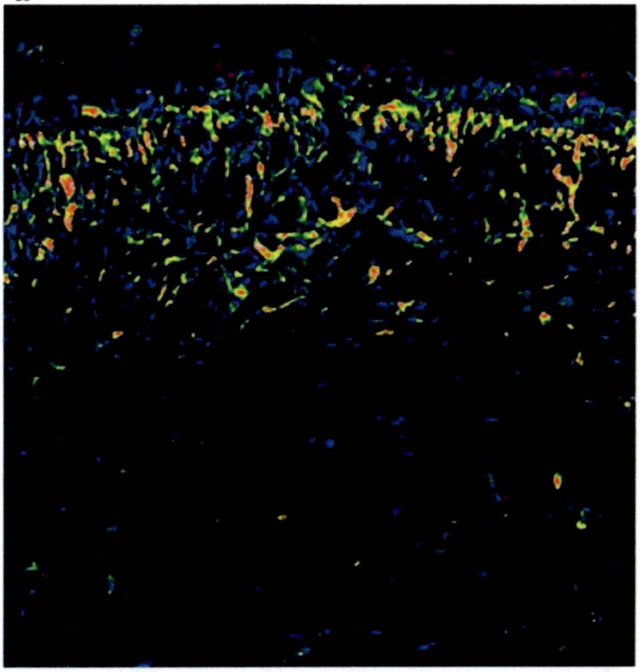

b

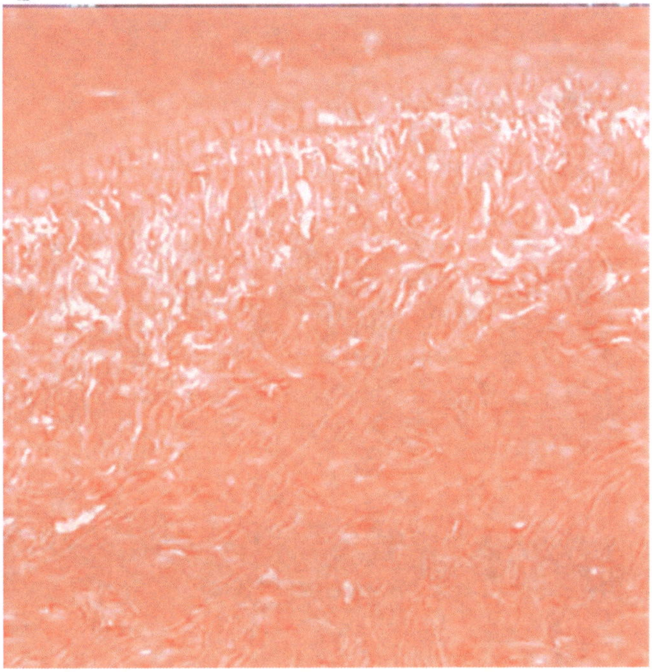

c

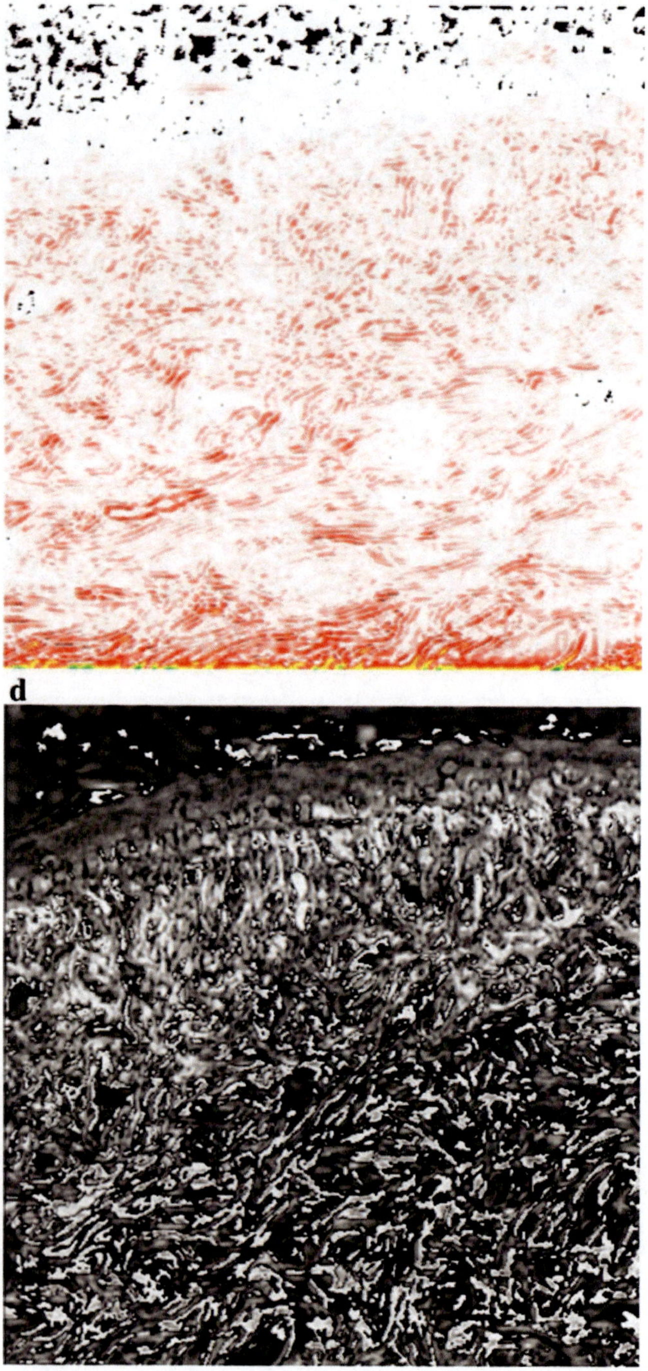

Fig. 10 (continued)

e

f

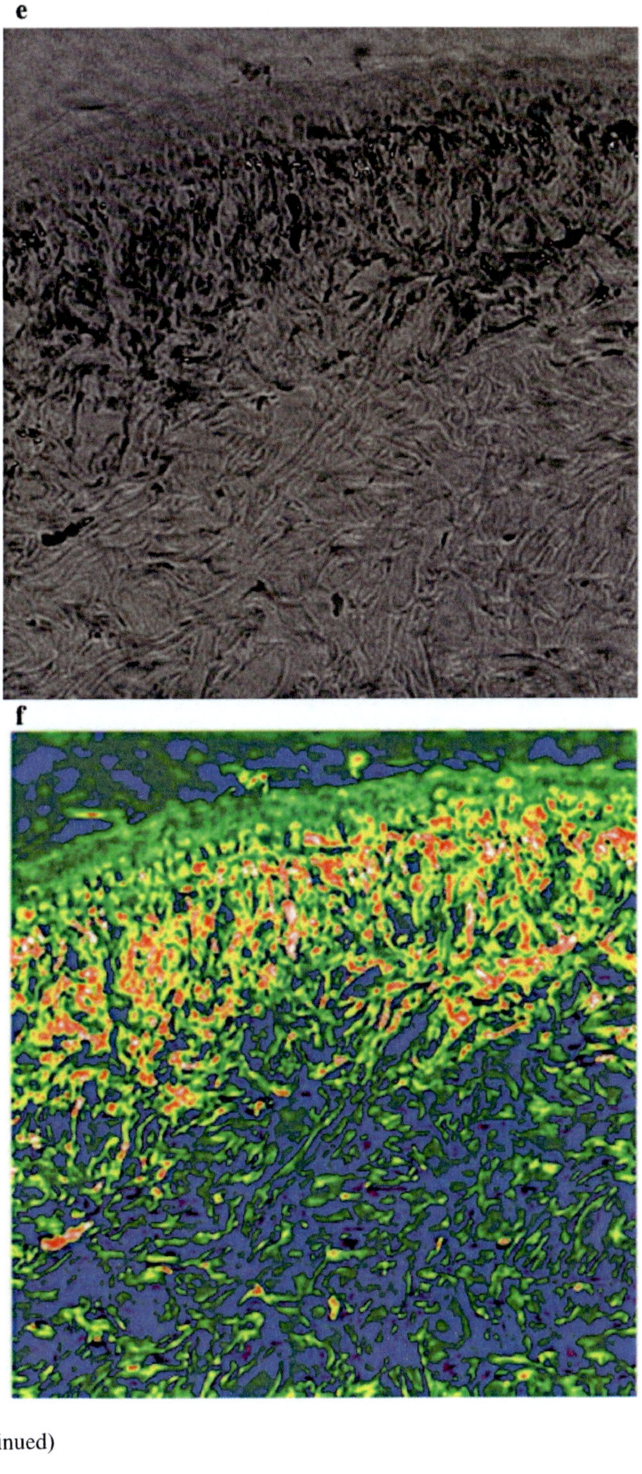

Fig. 10 (continued)

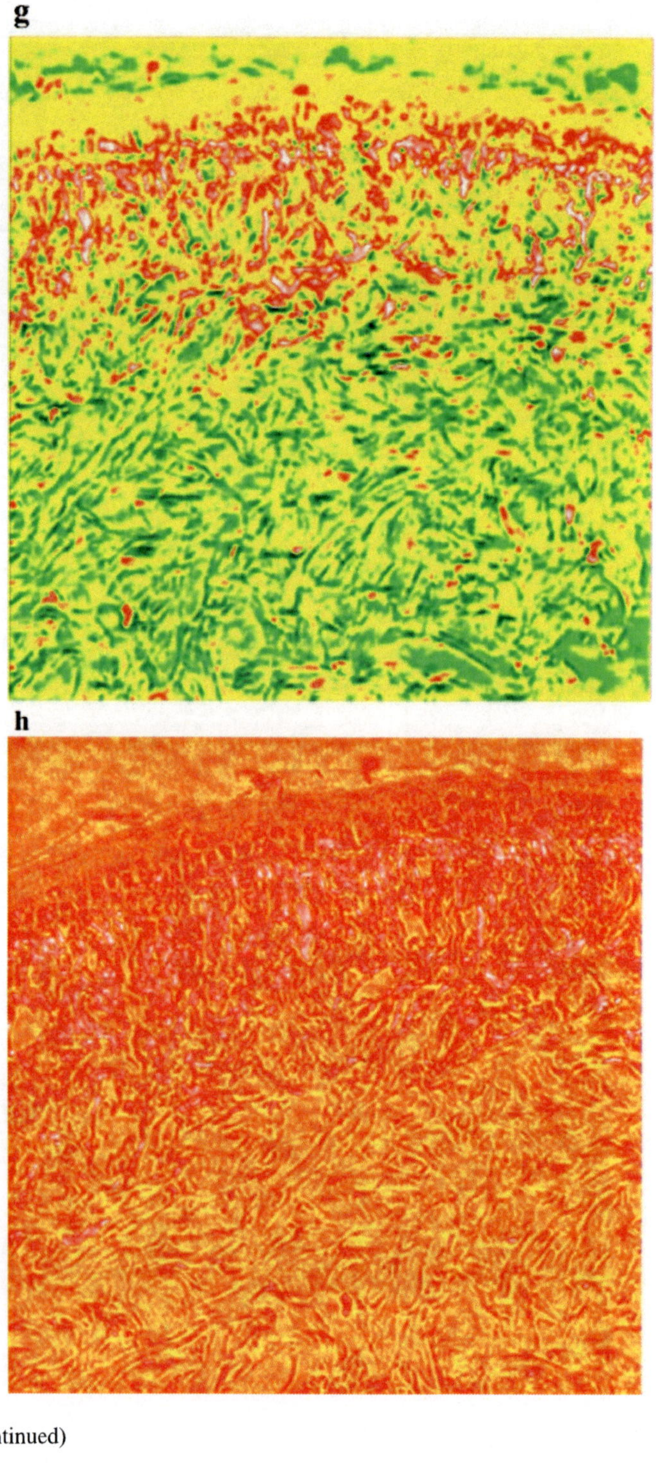

Fig. 10 (continued)

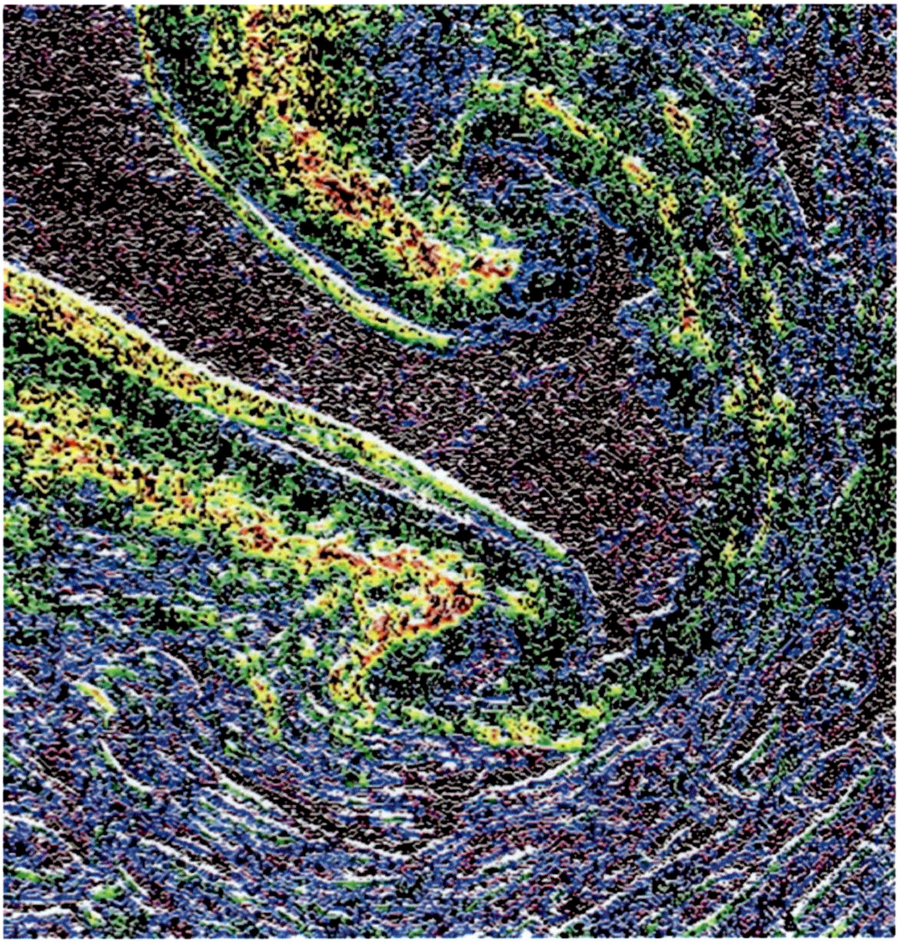

Fig. 11 The neural network analysis of the porous sensor region highlights the multi-layered structure of the sensor with a high activity at the location of the dendrites, whereas the external pore environment shows the lowest activity in dark-blue to black

a

b

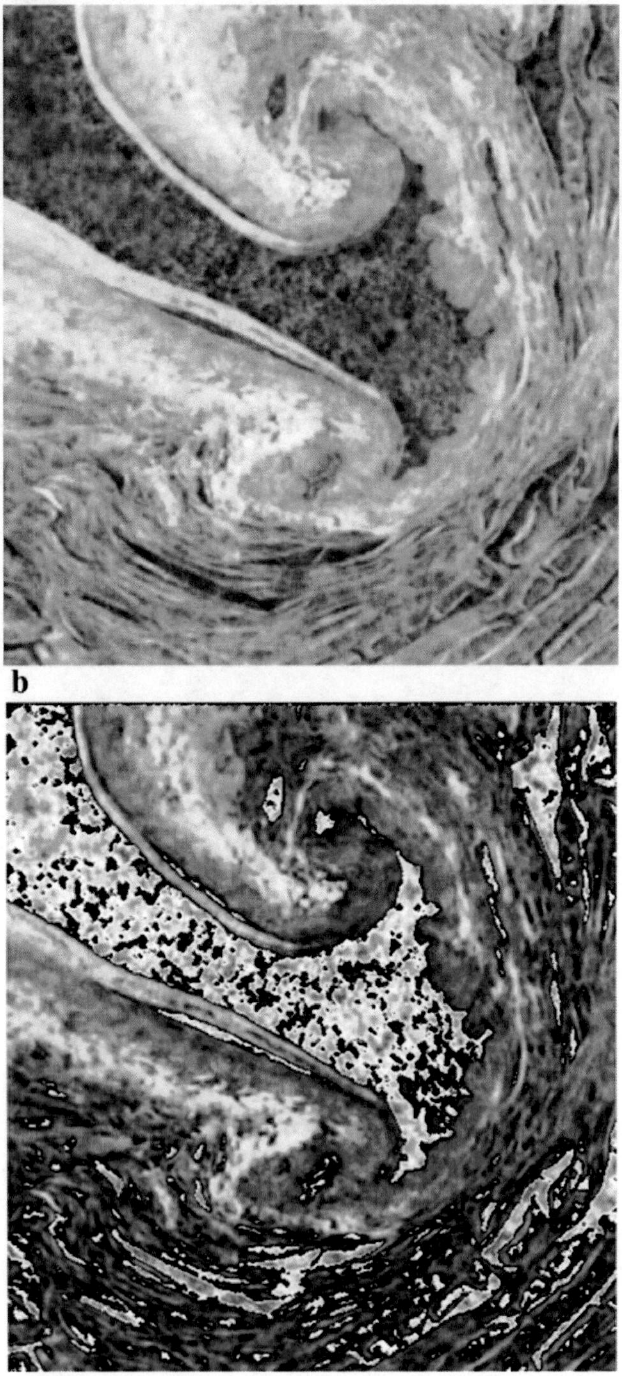

Fig. 12 **a** The neural network set the base for the analysis of the structure. The bright part shows the region of the lipoproteins of the dendrites and the darker region in the bottom of the figure was the location of the nerve endings. **b** Three-dimensional analysis of the porous structure reveals more structural contrast amongst the deep nerves seen primarily as dark areas in the bottom quarter of the figure, and less in the dendrite regions (light grey). **c** The neural network highlights the lipoproteins and dendrites (red) in contrast to the deeper nerve activity (purple)

c

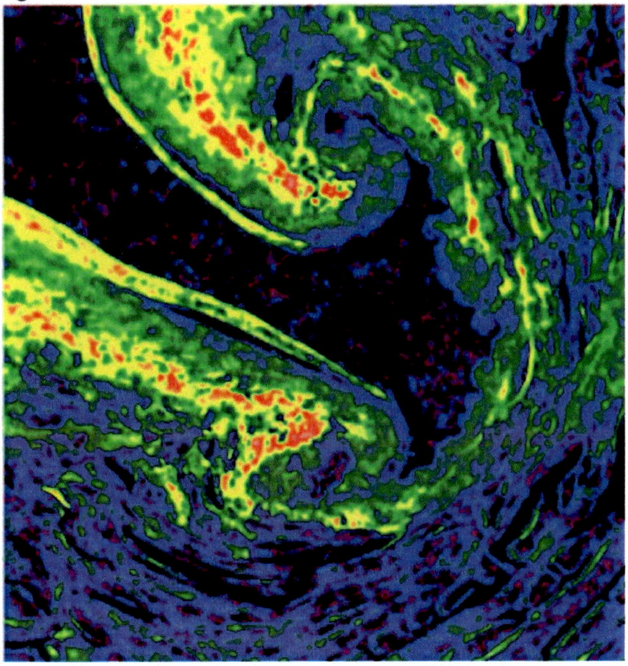

Fig. 12 (continued)

5 Conclusion

Infrared sensor regions of vipers, pythons and boas share similarities in respect to the presence of multi-layered lipoprotein-rich dendrite areas connected to nerve structures in the connective tissue. The sensory system of these snakes uses open end nerves (i.e. dendrites) as detectors. By neural network analysis, anatomical information of the sensor structure can be transferred towards functional analysis of the sensor by extracting areas of high neural activity and connectivity. Since the concentration of neural activity in this region is high [56], and the use of many sensors allow the system to overcome the background noise, the biological activity of the thermal sensor of the infrared sensor can be largely improved. The analysis and understanding of this principle may help in future approaches for development of biomimetic tools such as sensor-imitating microchips.

Acknowledgements We thank Carlos Ramirez for the preparation of the sections for this study. We are grateful to Mona F. Melhem, MD for help with the histology. The laser microdissection was performed under Dr. Tatiania Krasiva (LAMMP facility supported by the NIH under grant from the National Center for Research Resources # P41RR01192).

References

1. Wolken, J.J.: Light Detectors Photoreceptors, and Imaging Systems in Nature. Oxford University Press (1995)
2. Evans, G.: Nature **202**, 211 (1964)
3. Evans, G., Kuster, J.E.: Ecology **47**, 211–216 (1980)
4. Israelowitz, M., Rizvi, S.W., von Schroeder, H.P.: J. Lumin. **126**, 149 (2007)
5. Israelowitz, M., Kwon, K.A., Rizvi, S.W., Gille, C., von Schroeder, H.P.: J. Bionic Eng. **8**, 129 (2011)
6. Bullok, T.H., Cowles, R.B.: Science **115**, 541–543 (1953)
7. Gamow, I.R., Harris, J.F.: Sci. Am. **228**, 94–101 (1973)
8. Amemiya, F., Ushiki, T., Goris, C.S., Atobe, Y., Kusunoki, T.: Anatopical Rec. **246**, 135 (1996)
9. de Cock, B.: J. Neurosci. **29**, 7519 (1981)
10. Newman, E.A., Gruberd, E.R., Hartline, P.H.: J. Comp. Neurol. **191** (1981)
11. de Cock, B.: J. Comp. Physiol. **151**, 461 (1983)
12. Meszler, R.M.: J. Comp. Neurol. **220**, 229 (1983)
13. Waterman, T.H.: Animal Navigation (Scientific America Library. W. H. Freeman, New York (1980)
14. von Düring, M., Miller, M.R.: Sensory nerve endings of the skin and deeper structures. In: Gans, C., Northcutt, R.G., Ulinski, P. (eds.) Biology of the Reptilia, vol 9: Neurology A, pp 407–441. Academic, London (1979)
15. Amemiya, F., Goris, C.S., Masuda, R., Kishida, Y., Atobe, Y., Ishii, N., Kusunoki, T.: Biomed. Res. **16**, 411 (1995)
16. Campbell, A.L., de Cock, B., Stone, M.O., Church, D.O., Grase, M.S.: J. Struct. Biol. **126**, 105 (1999)
17. Klein, M.C., Deuschle, J., Gorb, S.N.: J. Comp. Physiol. **196**, 659 (2010)
18. Klein, M.C.G., Gorb, S.N.: R. Soc. Publ. (2012)
19. Chiasson, R.B., Bentley, D.L., Lowe, C.H.: Herpetologica **45**, 430 (1980)
20. Krochmal, A.R., Bakken, G.S., LaDuc, T.J.: J. Exp. Biol. **207**, 4231 (2004)
21. Gracheva, E.O., Ingolia, N.T., Kelly, Y.M., Cordero-Morales, J.F., Hollopeter, G., Chesler, A.T., Sánchez, E.E., Perez, J.C., Weissman, J.S., Julius, D.: Nature **464** (2010)
22. Berson, D.M., Hartline, P.H.: J. Neurosci. **8**, 1074 (1988)
23. Bullock, T.H.: Commun. Behav. Biol. A **1**, 10 (1968)
24. Bullock, T.H., Fox, W.: Q. J. Microsc. Sci. **98**, 219 (1957)
25. Hartline, P.H., Kass, L., Loop, M.S.: Science **199**, 1225 (1978)
26. Bleichmar, H., Robertis, D.E.: Z. Für Zellforsch. Mikrosk. Anat. **56**, 748 (1962)
27. Van Dyke, J.U., Grace, M.S.: Anim. Behav. **79**, 993 (2010)
28. Chiba, B.A., Menaker, M.: Vis. Neurosci. **16**, 909 (1999)
29. Chailapakula, O., Wonsawat, W., Siangprohb, W., Grudpan, K., Zhaod, Y., Zhud, Z.: Food Chem. **109**, 876 (2008)
30. Wilson, J.S.: Anat. Rec. **4**, 267 (1910)
31. Beaudoin, A.: J. Forensic Identif. **54**, 413 (2004)
32. Refat, N.A., Ibrahim, Z.S., Moustafa, G.G., Sakamoto, K.Q., Ishizuka, M., Fujita, S.: J. Biochem. Mol. Toxicol. **22**, 77 (2008)
33. Lillie, R.D.: Conn's Biological Stains. Williams & Wilkins, Baltimore, MD, USA (1969)
34. Lansink, A.G.W.: Histochemie **16**, 68 (1968)
35. Pfüller, U., Franz, H., Preiß, A.: Histochemistry **54**, 237 (1977)
36. Wootton, C.: A Practical Guide to Video and Audio Compression: From Sprockets and Rasters to Macroblocks. Elsevier (2005)
37. Baker, J.R.: Principles of Biological Microtechnique. Methuen, London (1970)
38. Orba, Y., Tanaka, S., Nishihara, H., Kawamura, N., Itoh, T., Shimizu, M., Sawa, H., Nagashima, K.: Cancer **99**, 198 (2003)

39. Emmert-Buck, B.R., Bonner, R.F., Smith, P.D., Chuaqui, R.F., Zhuang, Z., Goldstein, F., Weiss, R.A., Liotta, L.A.: Science **274**, 998 (1996)
40. Espina, V., Heiby, M., Pierobon, M., Liotta, L.A.: Expert Rev. Mol. Diagn. **7**, 647 (n.d.)
41. Curran, S., McKay, J.M., McLeod, H.L., Murray, G.I.: Mol. Pathol. **53**, 64 (2003)
42. Kihara, A.H., Moriscot, A.S., Ferreira, P.J., Hamassaki, D.E.: J. Neurosci. Methods **148**, 103 (n.d.)
43. Weinberger, R.: Practical Capillary Electrophoresis, 2nd ed. Academic Press (2000)
44. Marasco, P.L., Dereniak, E.L.: SPIE, pp. 363–378 (1993)
45. Holst, G.C.: Common Sense Approach to Thermal Imaging. JCD Publishing, Winter Park, FL, USA (2000)
46. Titterington, D.M.: Stat. Sci. **19**, 128 (2004)
47. Reuter, M., Bohlmann, S.: Izmir, Turkey (2011)
48. Reuter, M.: Waikoloa, Hawaii, USA (2008)
49. Kaski, S.: Acta Polythecnica Scand. Math. Comput. Manag. Publ. Finn. Acad. Technol. 57 (1997)
50. Duggery, R.G., Clarke, R.B.: Biochim. Biophys. Acta **1080**, 231 (1991)
51. Deka, L., Quddus, M.: Accid. Anal. Prev. **65**, 105 (2014)
52. Kotsiantis, S.B.: Informatica **31**, 249 (2007)
53. Bahiraei, M., Hosseinalipur, S., Zabihi, K., Taheran, E.: Adv. Mech. Eng. **2012** (2012)
54. Xinzhi, L.J.: Inf. Vis. **3**, 49 (2004)
55. Grace, M.S., Church, D.E., Kelly, C.T., Lynn, W., Cooper, T.M.: Biosensory Bioelectron. **14**, 53 (1999)
56. Israelowitz, M., Weyand, B., Leiterer, C., Munoz, V., Martinez-Tomas, C., Herraiz-Llacer, M., Slowik, I., Beleites, C., Fritzsche, W., Krafft, C., Henkel, T., Reuter, M., Rizvi, S., Gille, C., Reimers, K., Vogt, P., von Schroeder, H.P.: New J. Sci. (2014)
57. Boone, D.R., Sell, S.L., Hellmich, H.L.: J. Vis. Exp. **74**, e50308 (2013)
58. Dereniak, E.L., Boreman, G.D.: Infrared Detectors and Systems. Wiley (1996)
59. Frieden, B.R.: Probability, Statistical Optics and Data Testing. Springer, New York, NY, USA (1991)
60. van der Ziel, A.: Noise in Solid State Devices and Circuits. Wiley, New York, NY, USA (1986)
61. Israelowitz, M., Rizvi, S.W., Holm, C., Gille, C., von Schroeder, H.P.: J. Lumineces 149–153 (2007)
62. Weyand, B., Israelowitz, M., Henkel, T., Munoz, V., Reimers, K., Vogt, P., Rizvi, S.W., Gille, C., Holm, C., von Schroeder, H.P.: Elsevier (2013)
63. Israelowitz, M., Rizvi, S.W., Holm, C., Gille, C., von Schroeder, H.P.: J. Bio. Eng 129–139 (2011)
64. Kartalopoulos, S.V.: Understanding Neural Networks and Fuzzy Logic: Basic Concepts and Applications. The Institute of Electrical and Electronics Engineers, New York (1996)
65. Kasabov, N.K.: Foundations of Neural Networs and Fuzzy Systems and Kwoledge Engineering. MIT, Cambridge, MA (1996)
66. Goris, R.C., Atobe, Y., Nakano, M., Hisajima, T., Funakoshi, T., Kadota, T.: Microvasc. Res. **65**, 179 (2003)
67. Israelowitz, M., Rizvi, S.W., Karmer, J., von Schroeder, H.P.: Protein Eng. Des. Sel. **18**, 329 (2007)
68. Israelowitz, M., Weyand, B., Rizvi, S.W., Gille, C., von Schroeder, H.P.: Protein Modeling and Surface Folding by Limiting the Degree of Freedom. Springer, New York, Dordrecht, London (2013)

Biomimetics Applications

Optical Oxygen Measurements Within Cell Tissue Using Phosphorescent Microbeads and a Laser for Excitation

Elmar Schmälzlin, Mariel Nöhre, and Birgit Weyand

Abstract In the field of tissue engineering, spherical phosphorescent microprobes are well suited to quantify the oxygen content in close proximity to the cells during their growth phase. When using standard seed trays like glass dishes, probe excitation and signal acquisition can be realized by applying a fluorescence microscope. However, when using a bulky bioreactor a custom-built optics system with a laser as the excitation source had to be employed. Here we describe the basic principles of spherical optical oxygen microprobes for use in standard cell cultures and the system modifications with a laser, which were required in order to perform similar measurements from a distance in bioreactor cultures.

Keywords Optical oxygen measurements · Phosphorescent dyes · Microwell bioreactor · Differential laminar flow bioreactor

1 Introduction

Availability of oxygen strongly affects the metabolism of aerobic mammalian cells. Oxygen content influences growth, migration and differentiation of various cell types such as stem cells, neuronal cells, vascular cells or chondrocytes [1–3] and plays an important role in cell culture and growth in bioreactors [4]. Various techniques for monitoring the oxygen content in bioreactors are available [5], however, techniques which allow oxygen monitoring directly within cell tissue are rare.

To measure oxygen concentrations in aqueous solutions, traditional Clark electrodes, which measure the electric current caused by the chemical reduction of

E. Schmälzlin (✉)
Colibri Photonics GmbH, Friedrich-Engels-Str. 23, 14473 Potsdam, Germany
e-mail: es@colibri-photonics.com

M. Nöhre · B. Weyand
Hannover Medical School, Department of Plastic, Aesthetic, Hand and Reconstructive Surgery, Hannover, Germany

M. Israelowitz et al. (eds.), *Biomimetics and Bionic Applications with Clinical Applications*, Series in BioEngineering,
https://doi.org/10.1007/978-3-319-53214-1_8

oxygen, are still popular [6]. But with the successful development of high power light emitting diodes (LEDs), optical oxygen measurements based on phosphorescence quenching have become more significant.

Regardless of their function principle, oxygen sensors are mostly rod-shaped and intended to be immersed into the liquid under investigation, i.e. into a culture medium. These probes are technically well-suited for measuring oxygen within a stirred cell suspension. However, when growing 3D tissues, the applicability of standard designed submersible sensors is limited. The oxygen content of the supernatant culture medium gives little evidence about the actual oxygen content in the close proximity of the cells. Indeed, it is the micro-environmental oxygen concentration that is crucial for a natural development of the tissue [7].

One approach to determine oxygen contents within 3D tissue is the use of needle-type micro sensors. Needle-type oxygen microprobes are commercially available as miniaturized Clark electrodes [8] and as tapered fiber optic probes [9, 10]. With tip diameters down to 8 μm, and with the use of a micromanipulator, these sensors can be directly inserted into cell samples. But their small calibers and their delicate structures make their handling difficult and prevent a routine use in tissue bioreactors. Furthermore, the penetration procedure unavoidably damages the samples and may trigger oxygen-consuming biochemical stress reactions. External oxygen may diffuse through the puncture sites and affect the measurement results. In addition, needle-type sensors require electrical or optical connection wires, which prevent the use of these sensor types in closed systems, i.e. channels or compartments in microfluidic chips.

An interesting alternative technique is to stain cell tissue with oxygen-sensitive nanoparticles or soluble indicator dyes [11, 12]. Essentially, this method provides valuable information about the intracellular oxygen. To introduce the sensors into the cells, various delivery methods like gene-gun, electroporation or incorporation of the dye solution by the cells during incubation [13–15]. However, most of these techniques have generally been intended for use on the laboratory bench and can hardly be transferred to the interior of a bioreactor system. Additionally, soluble oxygen sensors harbor the risk of photochemical cell damage, due to emerging singlet oxygen [16] (see Sect. 3.2).

A practical approach is the use of particle-based optical phosphorescent microprobes [17]. Entrapped into a suitable host matrix, neither the dyes nor the emerging singlet oxygen have direct contact with the cells. Such microprobes can be manufactured small enough to provide information about the oxygen content in proximity to the cells. Using a laser, these microprobes can be easily excited from afar. In the following the function principle of optical oxygen microprobes and application examples of this technique in a laminar flow bioreactor are presented.

2 Model and Methods

2.1 Functional Principle of Optical Oxygen Measurements

Optical oxygen measurement methods are based on phosphorescent dyes. The phosphorescence of these dyes is quenched by ambient oxygen. Optical oxygen probes, corresponding measurement techniques, and biological applications which have been established in the past are summarized in [15, 18–20].

A general description of the different electronic states in which a sensor dye molecule can be is explained by the Jablonski diagram (Fig. 1).

After irradiation with light the electronic state of the molecules changes from the ground state (S_0) to a higher energetic excited singlet (S_1). Phosphorescent molecules can subsequently perform an electronic spin flip to reach an excited triplet state (T_1); this is a radiation-free process termed intersystem crossing. The dye molecule can return from the triplet state to the ground state by emitting a photon. Such a direct return is spin-forbidden, which means in practice that it happens with a time delay and the light emission shows a remarkable decay time in the range of micro- or milliseconds. Triplet emissions with such long decay times are termed as phosphorescence, whereas fast emissions on a nanosecond time scale are called fluorescence. Sensor molecules in the triplet state are prone to interact with molecular oxygen. During a collision, energy is transferred to the oxygen molecule, resulting in singlet oxygen ($^1O_2$) and a radiation-free return of the sensor molecule

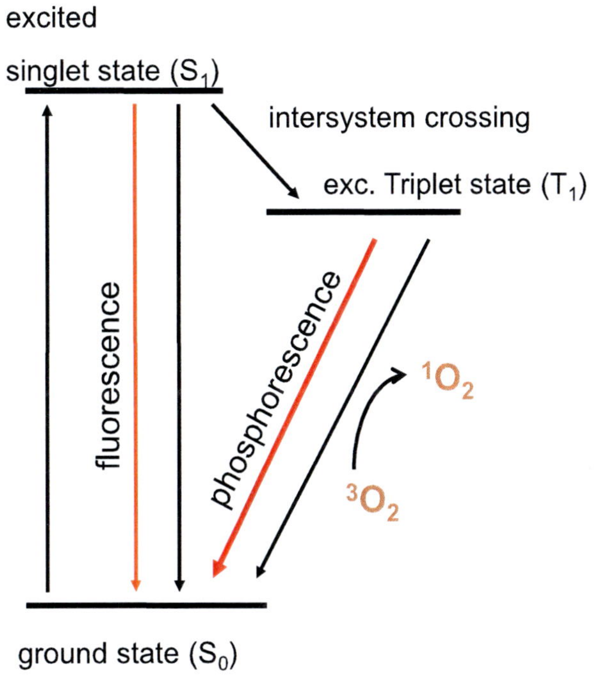

Fig. 1 Jablonski diagram of a phosphorescent oxygen sensor molecule

to its singlet ground state. As a result the phosphorescence intensity and the decay times decrease with increasing ambient oxygen concentration. This process is termed dynamic phosphorescence quenching. Due to the special electronic configuration of molecular oxygen, namely the existence of a triplet ground state ($^3O_2$), the energy transfer occurs very specifically.

Other gases like carbon dioxide and nitrogen as well as the pH value and ingredients of the culture media do not show evidence of interference with the quenching process. An exception is nitrogen monoxide (NO), which quenches the phosphorescence similar to oxygen. However, under native conditions the NO concentration is negligible compared to the oxygen content.

To have a feasible oxygen probe, the phosphorescent sensor molecules are embedded into a host polymer [19]. A great advantage of using host polymer-based oxygen sensors is that nearly all user-defined shapes with almost unlimited miniaturization potential can be manufactured, e.g. discs, spots, layers, coatings, microspheres. The polymer matrix enhances the phosphorescence intensity and shields the sensor molecule cross sensitivities. A sufficient extended matrix prevents cytotoxic singlet oxygen to reach the cell clusters. Using a polystyrene matrix with dimensions in the micrometer range, the generated singlet oxygen already reacts back within the matrix. Additionally, scavengers can be added, which catalyze a back reaction early enough to exclude any regression of singlet oxygen and additionally reduce the photo bleaching, i.e. the oxidation of sensor dye, to a far extent [21].

2.2 *Relation of Phosphorescence and Oxygen Content*

Dynamic phosphorescence quenching is described by the Stern-Volmer equation:

$$\frac{I}{I_0} = \frac{\tau}{\tau_0} = 1 + K_{SV} \cdot [O_2] \tag{1}$$

$[O_2]$ is the oxygen concentration. In the field of optical oxygen measurements, $[O_2]$ is frequently specified in % air saturation, shortly "% air", which is a physical meaningful unit for both, gaseous and aqueous environments. 100% air corresponds to 21% (v/v) or to air-saturated water (6.7 mg/L oxygen at 37 °C and 1013 hPa [22]), respectively. τ and I are the corresponding decay time and the signal intensity, respectively, and τ_0 and I_0 are the decay time and the signal intensity in the absence of oxygen.

K_{SV} is the quencher constant, which is related to the collision frequency of the molecules and therefore temperature-dependent.

For an ideal Stern Volmer behavior, two calibration points would be sufficient to determine K_{SV}. However, if solid probes are used, a deviation from Eq. (1) will occur. Due to microheterogeneities of the matrix, the individual sensor molecules are not equally accessible to oxygen molecules [23, 24]. Generally, sensor molecules that are located in the center of the matrix are less reachable for oxygen and therefore tend to show a higher average decay time than sensor molecules located near the surface. This behavior theoretically could be characterized with a series of quencher constants. In practice, it is sufficient to split the sensor molecules into two fractions: Assuming that one fraction f is quenched equally by oxygen and a second fraction $1-f$ is completely inaccessible for oxygen leads to a slightly modified Stern-Volmer equation [20].

$$\frac{I_0}{I} = \frac{\tau_0}{\tau} = \frac{f}{1 + K_{\mathrm{SVemp}} \cdot [\mathrm{O_2}]} + \frac{1-f}{1} \tag{2}$$

f is a value between 0 and 1, indicating the homogeneity of an oxygen probe with regards to the oxygen accessibility of the sensor molecules.

To determine the calibration curve of a real solid oxygen sensor by use of Eq. (2), three calibration points are necessary. f mainly depends on the polymer material, the geometry and the dimensions of the sensor and on the staining process. It is worth noting that dividing the sensor molecules into just two fractions is empirical. Subsequently, the value of $K_{\mathrm{SV\ emp}}$ evaluated by a three-point calibration is also empirical and has lost its direct relation to the actual collision frequency.

In comparison to the intensity, the decay time is preferred as a benchmark for the related oxygen content because it is far less affected by external factors. Contrary to intensities, decay times do not depend on the sensor concentration, the optical pathways, level of illumination or adsorption and scattering characteristics of the sample. Furthermore, the decay time is far less affected by photo bleaching.

2.3 Determination of the Decay Time

There are two fundamental methods to measure phosphorescence decay times: Time-resolved fluorometry and phase modulation technique [18].

Time-resolved fluorometry means that a fluorescent or phosphorescent chromophore is excited with light and the light emission following the excitation switch-off is recorded timely resolved. In the case of intense and slowly decaying phosphorescence signals, an oscilloscope may be used to record the decay curve. To measure fast decay curves from fluorescent dyes, photon counting is popular. This method records the time span between the excitation switch-off and the arrival of the first photon at the photo detector. Step-by-step a histogram is generated, which finally corresponds to the decay curve. However, photon counting is not the adequate method in the case of phosphorescent dyes, because their microseconds decay times limits the repetition rate to a few kHz. Since single photon counting records only one photon within a duty cycle, such a low repetition rate is not suitable to generate an evaluable histogram within a reasonable time frame. Therefore, multichannel scaling is the better technique to record the decay behavior of faint phosphorescent signals. Multichannel scalers capture many photons within one duty cycle timely resolved, which accelerates the generation of the histogram by some orders of magnitude. However, this requires demanding electronics and detectors with very short dead times. The advantage of a straightforward recording of phosphorescent decay curves is the easy elimination of short-lived fluorescence background signals. This can be done by gating, i.e. activating the photodetector not before a certain delay time, or even more simply by cutting the fluorescence-interfered range of the decay curves before evaluation. A simplified version of time resolved fluorometry is called rapid lifetime detection (RLD) [25]. After switching off the excitation light source, the photodetector is activated only twice for two equal time intervals. From the integrated light intensities measured during the respective time intervals, an apparent decay time is calculated. However, RLD is significantly affected by constant levels of background light. Generally, due to the microheterogeneity of a solid state sensor, the resulting decay curves need stretched exponential analysis [24]. That means that the decay behavior has to be described by an exponential equation containing an additional stretching factor β, describing the microheterogeneity. However, this additional value complicates a definite correlation of the evaluated parameters with the actual oxygen concentration.

Not least because of the easy analysis, the technique of phase modulation is common for commercial optical oxygen meters. Phase modulation means to compare the phasing of excitation and emission signals [26]. The intensity of the excitation light is sinus-modulated. The time-delay of the phosphorescence emission leads to a phase shift in comparison to the excitation signal. Figure 2 shows the mode of operation. The time delay and hence the phase shift depends on the ambient oxygen concentration.

For the case of a single exponential decay, phase shift $\Delta\phi$ and decay time τ are related by

$$\tau = \frac{\tan \Delta\phi}{\omega} \tag{3}$$

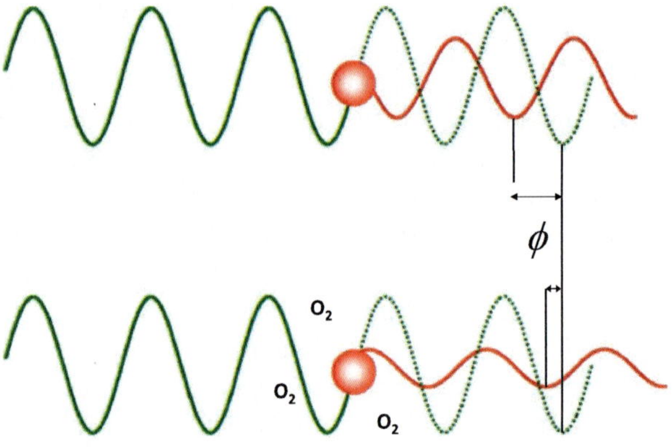

Fig. 2 Functional principle phase modulation measurements. An oxygen sensor is excited by sinusoidal intensity-modulated green light. The phosphorescence emission is time-delayed depending on the ambient oxygen content, which is reflected by the phase shift with regard to the excitation signal

where ω is the modulation frequency in radian measure. The adequacy of the modulation frequency depends on the decay time.

For an oxygen probes with a decay time ranging from 19 to 60 µs, a maximum phase shift change can be achieved at 6 kHz modulation frequency. The rule of thumb is to choose a modulation frequency that leads to a phase shift of 45° at an average decay time. Apart from the high measurement and analysis speed, it is a benefit that phase modulation takes only one parameter, namely the phase shift, which can be directly related to the oxygen content. Furthermore, unmodulated background light does not interfere with the phase shift. For real sensors showing a more complex decay behavior, Eq. (3) provides an average decay time $\overline{\tau}$, which is still conveniently suitable as a calibration benchmark using Eq. (2).

However, the standard phase modulation technique is only error-free, if there is no significant background fluorescence in the spectral range of the sensor signal. For large sensors showing high signal intensities the relative contribution of background fluorescence is small and therefore can be neglected in the majority of cases. However, when using faint microsensors, the fraction of fluorescence becomes significant. Background fluorescence is modulated with the same frequency as the sensor signal, but not phase shifted. The resulting superposition leads to a total signal with a reduced phase shift. Figure 3 shows the situation.

In order to circumvent this problem, a two-frequency phase modulation technique can be applied [27]. The excitation light is not only modulated with one, but with two superposed sinus frequencies ω_1 and ω_2. Since there are two modulation frequencies, two apparent phase shifts $\Delta\phi_{\mathrm{app},\omega1}$ and $\Delta\phi_{\mathrm{app},\omega2}$ are measured. From $\Delta\phi_{\mathrm{app},\omega1}$ and $\Delta\phi_{\mathrm{app},\omega2}$ the proportion of background fluorescence is being calculated and subsequently mathematically eliminated. Finally, the actual undistorted decay time of the sensor is received. This method is based on the assumption that fluorescence signals are always in-phase with the excitation light. According to Eq. (3) short fluorescence decay times (a few nanoseconds only) generate almost no phase shifts if modulations frequencies applied are within the kilohertz range.

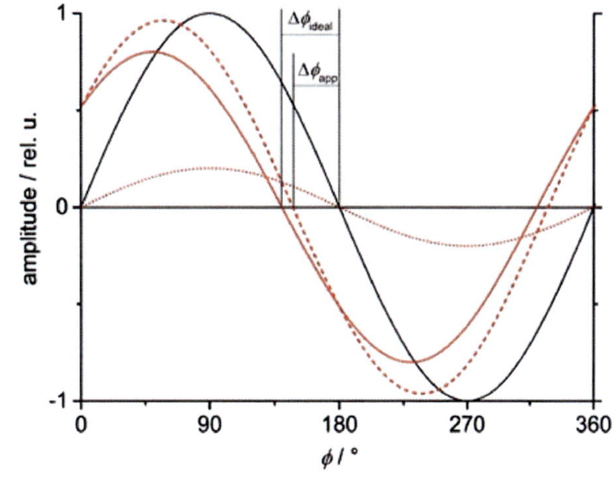

Fig. 3 Phase modulation in the presence of background fluorescence. The sensor phosphoresence (red solid line) is superposed by background fluorescence (red dotted line) leading to a total signal (red dashed line) with a reduced phase shift $\Delta\phi_{\mathrm{app}} < \Delta\phi_{\mathrm{ideal}}$ with regard to the excitation light (black solid line) (reprinted from Ref. [28])

2.4 Realization of an Optoelectronic System That Uses Two-Frequency Phase Modulation

The realization of the two-frequency phase modulation technique with regard to a commercially available system was performed by the Colibri Photonics GmbH. The system, called "OPAL" was described previously [28]. It was initially developed to be linked to common fluorescence microscopes (Fig. 4).

Excitation of the sensors is performed via a high power LED source, mounted at the fluorescence lamp port of the microscope. To avoid the photochemical side reactions, a green LED was used for excitation, even though violet or blue light was better matched to the absorption maxima of the sensor dye (see Sect. 2.5). The sensor signal was detected by a photomultiplier unit which was mounted at the camera port. The core of the OPAL system consisted of an optoelectronic control box including frequency synthesizer, a LED driver, lock-in amplifiers, and an USB interface for data transfer to a PC or a laptop. Figure 5 shows the technical structure of the hardware unit. The excitation light was modulated by two superposed sinus frequencies. The lock-in amplifiers allowed measurement of the respective phase shifts of the detector signal at each of the frequencies. The subsequent calculation of the background proportion and the evaluation of the oxygen content were done by software.

The system is modular, so if necessary, sensors, light source and detector can be exchanged with custom parts. As oxygen sensors, spherical microsensors called CPOx-beads are provided by default (Table 1), which are intended to be added to cell samples. Nevertheless, the system is generally compatible to all kind of decay

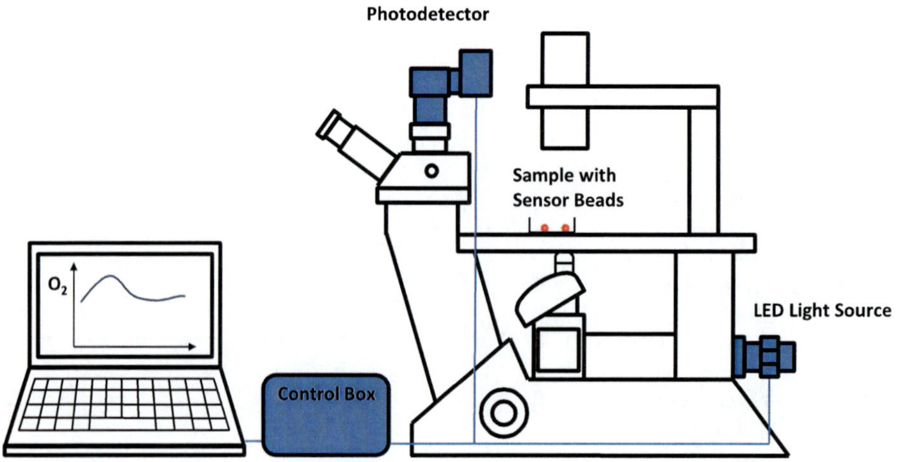

Fig. 4 Standard configuration of the OPAL oxygen measurement system with fluorescence microscope

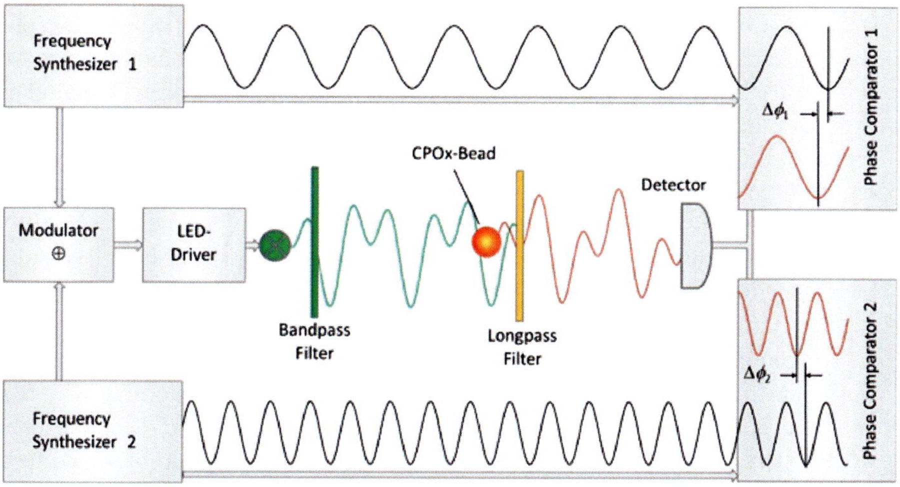

Fig. 5 Schematic structure of the control box of the optical oxygen measurement system "OPAL", which implies the two frequency phase modulation technique (reprinted from Ref. [28])

Table 1 Properties of the commercially available spherical oxygen microsensors Ru-CPOx and Pt-CPOx

Name	Matrix material	Diameter (μm)	Dye	Ex_{max} (nm)	Em_{max} (nm)	τ_0/μs @ 37 (°C)	τ_{air}/μs @37 (° C)	f
Ru-CPOx	crosslinked polystyrene	50	Ru $(dpp)_3$	440	600	5.3	4.0	0.70
Pt-CPOx	crosslinked polystyrene	50	PtTPFPP	390, 505, 540	650	60	19	0.77

time-based phosphorescent oxygen sensors. These spherical microsensors are intended to be used for monitoring the oxygen content within cell cultures that grow in well-plates, dishes and in channels and compartments of microbioreactors.

2.5 *Examples for Oxygen Sensors*

Many types of phosphorescent oxygen sensors have been described [18]. Most widespread sensor dyes are metal-organic complexes with heavy atoms. Fluorinated Pt-porphyrins and Ru-phenanthrolines are most popular. These compounds are commercially available due to their use as photosensitizers in the production of organic light emitting diodes and show high photo stability [29, 30]. When selecting a suitable matrix, oxygen solubility, the diffusion constant and also photo stability are important [19]. Additionally, the host matrix should provide for a sufficiently low lifetime for singlet oxygen to avoid singlet-oxygen accumulation [31]. For oxygen measurements within a physiological range, polystyrene has proved itself in praxis. Figure 6 shows the spectra of 50 μm diameter polystyrene sensor beads stained with Ruthenium-tris(4,7-diphenyl-1,10-phenanthroline) perchlorate (Ru(dpp)$_3$) and Pt(II) meso-Tetra(pentafluorophenyl)porphyrin (PtTPFPP), respectively. Table 1 shows the corresponding phosphorescence lifetimes and the homogeneity parameters *f*. The listed values are approximate depending on the exact manufacturing procedure.

To manufacture spherical oxygen microprobes, cross-linked polystyrene beads were immersed into a suitable organic solvent containing the senor dye. The beads swoll and absorbed the dye. After drying and washing, orange and red microspheres were obtained, respectively, which were biocompatible and could be added to cell cultures. An optimized staining and washing process allowed the production of uniform beads concerning their decay time-based calibration curves. The standard deviation of single beads with respect to their response to air saturation was less than 5%. Thus, depending on the demanded accuracy, only a single bead might be sufficient to measure oxygen quantitatively. Figure 7 shows a CPOx bead added to a culture of myeloma cells. For measurements within a physiologically relevant oxygen range, both, polystyrene beads with Ru(dpp)$_3$ and PtTPFPP are quite equally suitable. Due to its longer emission wavelength, the signal of PtTPFPP usually is somewhat less affected by shorter-wavelength background fluorescence. However, this is not the case in the presence of heme-related biomolecules, such as chlorophyll or hemoglobin. These compounds also have porphyrin structures and therefore also emit around 650 nm. In this case, Ru(dpp)$_3$ is more appropriate. Furthermore, an emission at a shorter wavelength is advantageous when using a photomultiplier as detector. Usually, the sensitivity of photo multipliers significantly decreases from the orange to the red spectral range.

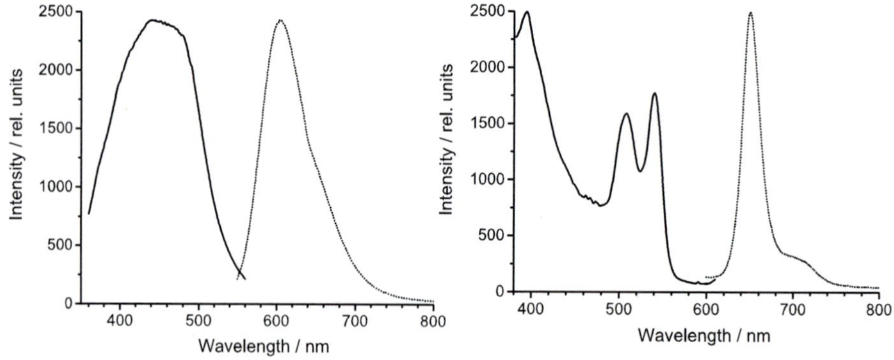

Fig. 6 Phosphorescence spectra of 50 μm polystyrene oxygen sensor beads stained with Ru (dpp)$_3$ (left) and PtTPFPP (right). Solid line: excitation spectrum; dotted line: emission spectrum

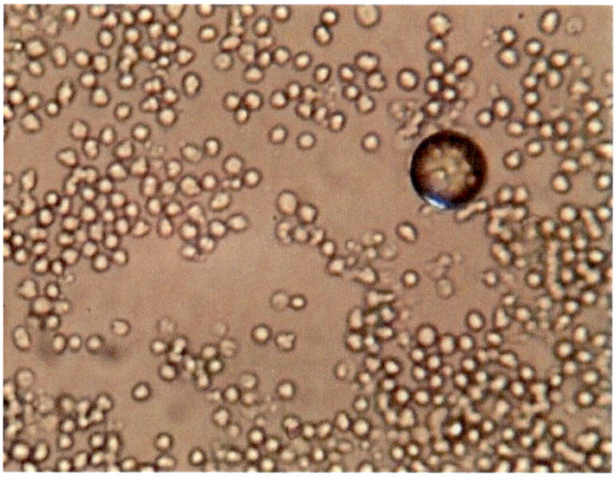

Fig. 7 CPOx bead spherical oxygen sensor in a myeloma cell culture. The diameter of the sensor sphere is 50 μm

For intracellular oxygen measurement, cell-penetrating nanoparticle-based sensors were produced [12]. These nanoparticles consist of a positive charged, cell penetrating polymer, which was originally developed for drug delivery systems. To receive an intracellular oxygen sensor, these nanoparticles were stained with PtTPFPP. The sensor is self-loading, i.e. an addition of a certain amount of sensor to the culture medium for 1 day with a washing step afterwards is sufficient to perform an experiment. However, there is no spacious matrix with photo stabilizers and therefore emerging singlet oxygen is not trapped effectively. This makes experiments with cells with soluble oxygen sensor challenging. To prevent cells from damage caused by the singlet oxygen, exposure times need to be as short as possible (see Sect. 3.2).

3 Examples of Biological Applications

3.1 Toxicity Tests in a Microwell Bioreactor with Liver Cell Spheroids

To perform toxicity tests, a so-called liver on a chip microbioreactor was developed [32, 33]. In a microwell insert, spheroids of liver cells were bred. Figure 8 shows the microbioreactor.

Via *microchannels* culture medium and chemicals could be added. The outer diameter of the reactor housing was 50.8 cm and could easily be placed at the stage of an inverted fluorescence microscope. The oxygen consumption of the liver spheroids, and hence their vitality, was indicated by the phosphorescence signal of added Ru-CPOx sensor beads. Figure 9 shows a liver cell culture with added sensor beads. Live dead staining after 28 days of perfusion proved the biocompatibility of

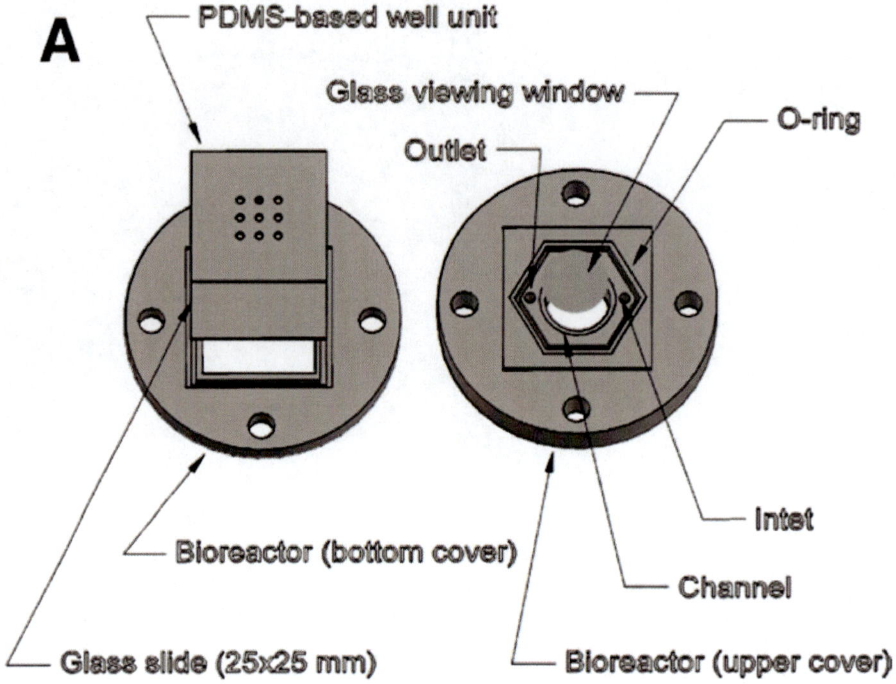

Fig. 8 Liver on a chip microbioreactor (reprinted from Ref. [32])

Fig. 9 Liver cells with orange oxygen sensors beads. Cells are stained ton identify living (green) and death (red) ones. Bar: 100 μm (reprinted from Ref. [32])

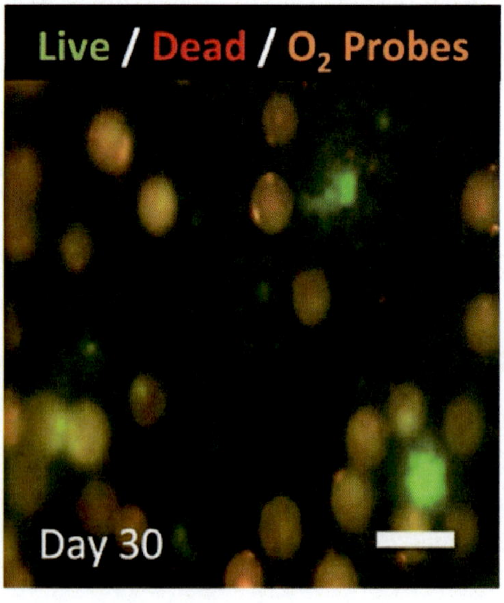

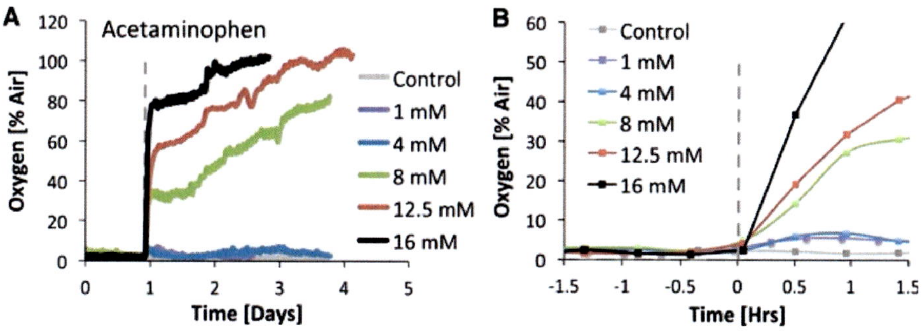

Fig. 10 **a** Dose-dependent loss in oxygen uptake of liver cells after addition of acetaminophen. **b** Close-up to show the fast response to all concentrations of acetaminophen (reprinted from Ref. [32])

the sensor beads: 98% viability and no accumulation of dead cells in the ambience of the sensor beads were found.

To perform the experiments the OPAL oxygen measurement system (see Sect. 2.3) was linked to an Olympus IX81 inverted fluorescence microscope with a climate control chamber. Due to the decay time-based measurement method, the results were independent of the optical focus and the numbers of sensor beads in the image field. The liver cells were exposed to increasing concentrations acetaminophen (paracetamol). Figure 10 shows the responses.

An immediate dose-dependent loss of oxygen update and thereafter a slow dose-independent loss of oxygen were observed. This two-stage response was in accordance with the corresponding interference of the mitochondrial electron transfer system. From the oxygen characteristics a TC_{50} value of 12.3 mM was received. The chip system was intended to replace animal experiments in the future. It was a major advantage that the condition of the liver cells could be monitored continuously, whereas animal dissections only provided information at a definite point of time.

3.2 *Intracellular Oxygen Measurements*

Nanoparticle-based oxygen sensors described in [12] (see also Sect. 2.5) are commercially available ("MitoXpress"[1], Agilent, Santa Clara United Stated). Initially, they were developed to be used with fluorescence lifetime imaging (FLIM) microscopes and fluorescence plate readers that are capable to quantify microseconds decay times. However, FLIM microscopes for a phosphorescence time scale are expensive and rare. Therefore, it was tested if the comparatively low-cost OPAL

[1]https://www.agilent.com/en/product/cell-analysis/real-time-cell-metabolic-analysis/plate-reader-metabolic-assays/mitoxpress-intra-intracellular-oxygen-assay-740893

oxygen measurement system is suitable to work with these intracellular sensors [34]. For this purpose, murine myeloma cells were stained with the nanoparticle-based oxygen sensor by incubating them over night in a nanoparticle solution. After washing with fresh culture medium, the well plate with the cells was placed on the stage of an inverted Olympus IX81 fluorescence microscope. The microscope was linked to the OPAL system and equipped with a climate control chamber. To induce changes of the intracellular oxygen content, the cells were treated with carbonyl cyanide-4-(trifluoro-methoxy)phenylhydrazone (FCCP) and antimycin A. FCCP uncoupled the respiratory chain leading to an increased oxygen consumption. Antimycin A inhibited the respiratory chain leading to apoptosis and thus to disruption of any oxygen consumption. Figure 11 shows the results.

Undisturbed cells showed about 76% air saturation internal oxygen content. After addition of FCCP the oxygen concentration dropped to 71% due to the increased respiration of the cells. After addition of antimycin A, the cells stopped their respiration and the oxygen concentration rose until 100% air saturation was reached. This result matched the behavior published in [12] using a time-resolved fluorescence plate reader. Thus, the OPAL was compatible to nanoparticle-based sensors for intracellular oxygen measurements. It is worth noting, that it was necessary to protect the cells from light as far as possible. The excitation light of the oxygen measurement system was only switched on for 7 s per measurement point. Loading a cell with an oxygen sensor dye, e.g. PtTPFPP, meaned that it is loaded with a dye for photodynamic therapy [35]. The singlet oxygen unavoidably arising during the quenching process damages the cells within a few minutes, seen as swelling and lack of characteristic responses to the metabolic stimuli. In contrast to

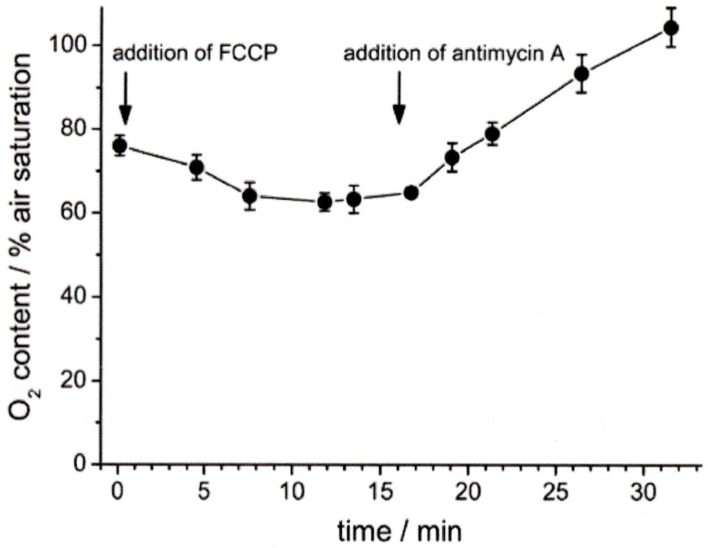

Fig. 11 Change of the intracellular oxygen content of myeloma cells after adding FCCP and antimycin A (reprinted from Ref. [34])

the larger spherical microsensors the unextended polymer matrix of the nanosensors did not completely prevent the emergence of singlet oxygen, which complicated intracellular oxygen measurements.

3.3 Oxygen Measurements with a Laser from a Distance in a Perfusion Bioreactor

The OPAL oxygen measurement system was implemented into a laminar flow bioreactor, which was intended for the breeding of three-dimensional bone tissue from stem cells [36]. Figure 12 shows a schematic drawing of the reactor and the optical components for the oxygen measurement. Figure 13, left shows a picture of the actual setup. The bioreactor was completely filled with culture medium. A cylindrical collagen-elastine-based cell carrier (Fig. 13, right) was located in a holder in the center the reactor. The specific cell differentiation was supported by flow-induced shear forces. Considering the liquid pressure above and below, bypass canals in the cell carrier holder could be opened and closed to align the laminar

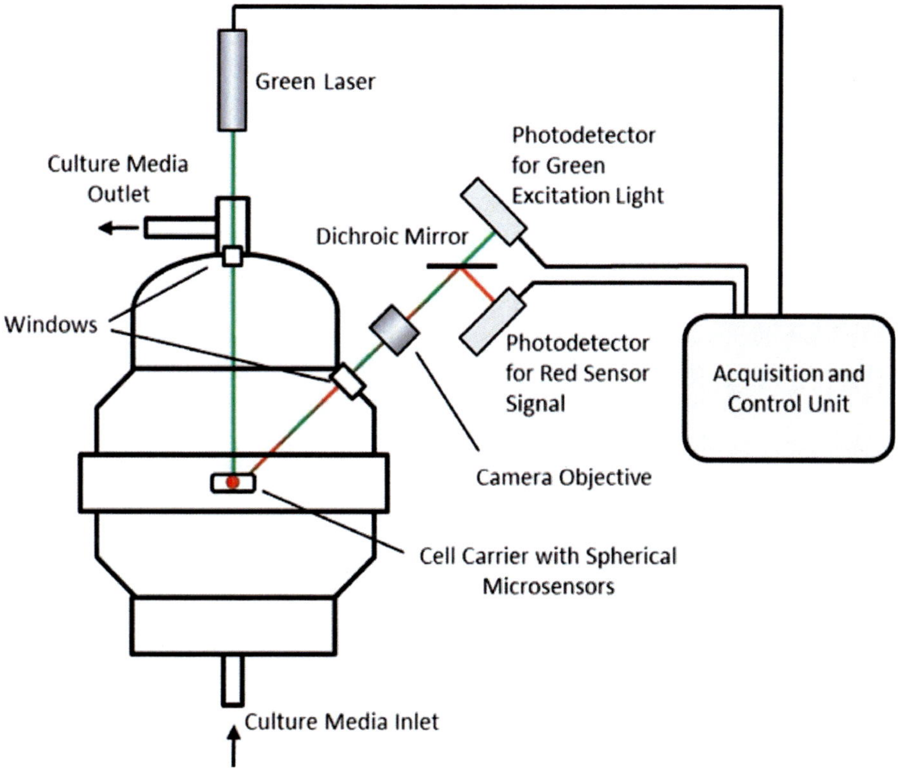

Fig. 12 Laminar flow cell bioreactor with an optical oxygen measurement system

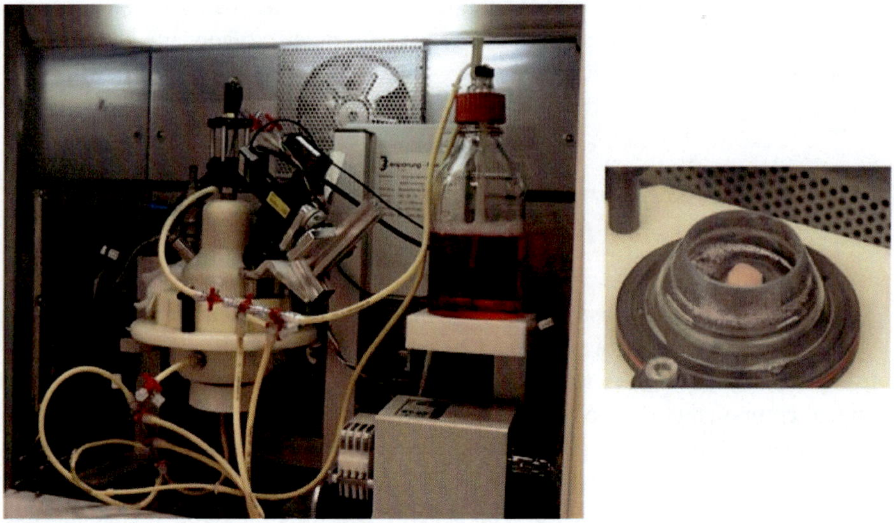

Fig. 13 Left: Laminar flow cell bioreactor with optical oxygen measurement system. Right: Collagen-elastin cell carrier inside the holder unit. The holder has to be fixed in the center of the bioreactor

flow. To measure the oxygen content within the cell culture, the carrier was cut and some Pt-CPOx beads were placed centrally inside the carrier. To avoid any interference with the controlled flow conditions inside the reactor, no further installations inside the reactor, i.e. light guides, LEDs etc., were allowed. Thus, all components of the oxygen measurement system had to be placed outside and the signal detection as well as the sensor excitation had to be realized from a distance. To guide sufficient excitation light to the cell culture a laser was used as light source. In contrast to the light of a LED a collimated laser beam could pass the distance through the culture media and reach the cell culture nearly loss free. The laser beam entered the reactor via a small glass window, passed through the culture medium for approximately 20 cm, and finally encountered the central cell carrier. A lens in front of the laser allowed an alignment of the beam diameter with regard to the size of the light spot at the position of the cell carrier and the entrance window. A second window built into the wall of the reactor was used to detect the phosphorescence signals of the CPOx beads. On a rail system in front of the window, the photo detector unit was aligned. The photo detector unit included a camera objective and two photomultiplier modules (H10723-20, Hamamatsu, Japan) with suitable green and red band pass filters in front of the respective active areas. A beam splitter cube separated the scattered green excitation light and the red sensor signal arising from the CPOx-beads. As shown below in Fig. 15 the introduction additional of a further detection channel for the green excitation light was necessary to eliminate phase deviations of certain laser types.

For selecting an appropriate laser source, wavelength and modulation capability had to be considered. If an intense laser beam passed through the culture medium,

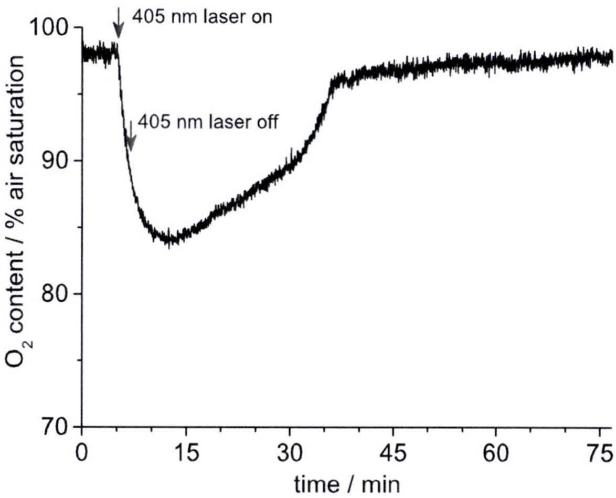

Fig. 14 Effect of violet laser illumination of the measured oxygen content in a cuvette. Violet light generates singlet oxygen, which does no more act as quencher

singlet oxygen might be generated photochemically, most likely catalyzed by riboflavin, which was a permanent ingredient of culture media [37, 38]. To demonstrate the possible interference, a 10 x 10 mm cuvette was filled with 2 mL culture medium (DMEM W/GLUTAMAX-I, PYR, 4.5G GLU; Invitrogen Nr. 31966021). The experiment was performed in combination with an inverted fluorescence microscope that was connected to the optical oxygen measurement system (OPAL, see Sect. 2.4). At the transparent cuvette bottom some Pt-CPOx beads were arranged. The cuvette was placed on the microscope stage and the oxygen measurement was started. As long as the cuvette was only irradiated with the green excitation light (530 nm) of the oxygen measurement system, an oxygen content of 100% air saturation was detected. Then the cuvette was additionally irradiated with a violet diode laser. As soon as the violet laser beam (405 nm, 3 mm beam diameter, ca. 10 mW, position of the beam ca. 1 cm over the cuvette bottom) passed the cuvette, the displayed oxygen concentration started to decrease. Figure 14 shows the apparent oxygen decrease.

The violet laser light induced the photochemical conversion of triplet oxygen to singlet oxygen and the displayed oxygen value decreased. Even after switching off the laser, the displayed oxygen content kept decreasing for the next 5 min. Since the lifetime of singlet oxygen in water was too short (4 μs [39]) for an accumulation, the sustaining decrease indicated that singlet oxygen reacted with ingredients of the culture medium and thus was permanently withdrawn from the solution. Since the sample was not stirred the oxygen-depletion had to be compensated by slow diffusion from the liquid surface. After 5 min, the oxygen concentration increased, until air saturation was reached again. The extended generation of singlet oxygen does not only affect the measurement results, but also shows cytotoxicity [37]. Therefore, violet and blue lasers were excluded and a green laser was chosen. Its wavelength is beyond the absorption spectrum of riboflavin but still adequate to excite the PtTPFPP oxygen sensor dye (Fig. 6, right.). The generation of singlet could also be induced by blue or violet light of high power LEDs. For this reason the OPAL oxygen measurement system was equipped with a green LED, even if green light did not fully match the absorption spectra of the sensor dyes.

For green laser light, 532 nm frequency doubled Nd:YAG solid state are most widespread and commercial available in many designs and price ranges. However, frequency doubled solid state lasers show on principle no simple linear dependency from the electrical driver current as it is the case for LEDs and semiconductor diode lasers. Thus, it is more difficult to receive a well-defined analog intensity modulation. However, a phase-stable analog modulation of the excitation light is a precondition for reliable decay time measurements via phase modulation technique. Even with electronic corrections the phase stability is unsatisfying, leading to signal variances and drifts.

To demonstrate this interference, the green high power LED of the OPAL oxygen measurement system was replaced by a Nd:YAG laser (50 mW, MGL-532-50, CNI, China). The analog modulation voltage was provided by an output of the OPAL hardware unit. The laser light was guided from the laser head to the lamp of the microscope via an 1 mm optical polymer fiber. On the microscope stage CPOx beads in air saturated water were placed. Figure 15 (0–400 s, left part of the curve) shows the resulting measurement curve.

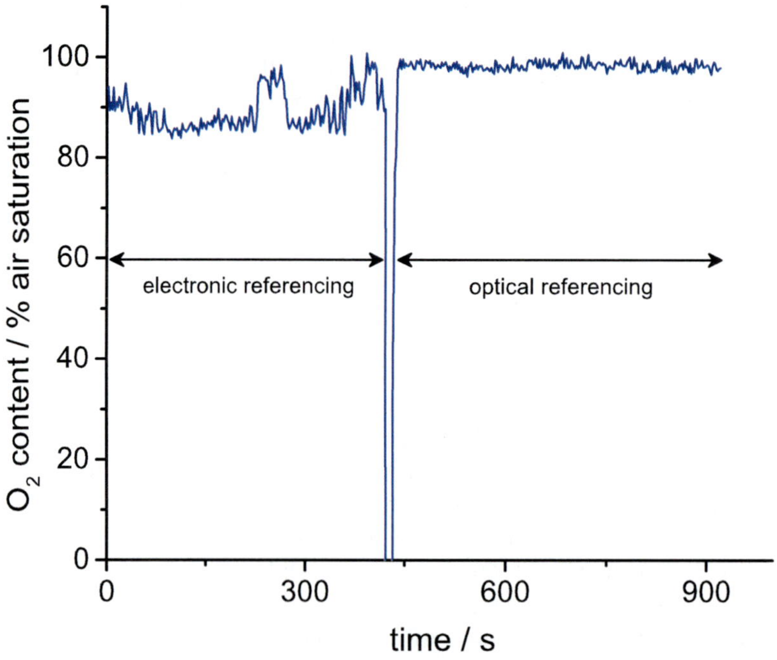

Fig. 15 Optical oxygen measurements using a 532 nm frequency-doubled solid state laser as intensity-modulated excitation source. Left part (electronic referencing; standard): The oxygen-dependent phase shifts of are determined with regard to the control voltage of the laser. The laser shows phase drifts and phase hopping leading to inaccurate results. Right part (optical referencing): The phase shifts are determined with regard to the actual excitation light, which was additionally monitored with a second photo detector. The noise of the curve is significantly reduced

The curve was noisy and showed drift and hopping, preventing accurate measurements. In the standard configuration, the OPAL system uses the LED driver modulation voltage as internal reference for determining the phasing. Obviously, in the case of a solid state laser, the phasing of the modulation voltage was not in any stable relation to the actual light output. However, the situation could be considerably improved by changing to optical referencing. For this purpose, a second channel was added to the hardware unit of the OPAL measurement system. A second photomultiplier measured the green excitation light continuously. If the actual phasing of both, excitation light and sensor signal, were measured simultaneously, the interference of phase hopping and phase drifts was eliminated by simple subtraction. Figure 15 (>450 s, right part of the curve) shows the benefit of the optical referencing. After applying the optical reference channel the curve proceeded considerably smoother. Hopping and drift were completely vanished.

As an alternative of a frequency doubled Nd:YAG solid state laser, green laser diodes modules with wavelengths around 515 nm can be used. It is an advantage that such diode lasers can be phase-stably modulated like LEDs by simply controlling the current. For comparative purposes, a 515 nm, 25 mW diode laser module (FWML-515-25-AM, Frankfurt Laser Company, Germany) with an analog modulation voltage input was used.

Both, the green solid state laser and the green diode laser, were used alternatively for test measurements at the bioreactor. Generally, when guiding a laser beam through several centimeters of culture medium, consideration has to be given to absorbance, scattering, and fluorescence that might arise from colored substances, e.g. pH indicators. In fact, the two frequency phase modulation techniques fades out the interference of background fluorescence, but only to a certain extent. If the background fluorescence overimposes the sensor signal, the fade out will become imperfect leading to deviations of the measurement results. Consequently, avoiding background fluorescence is still an important issue. Fortunately, the used pH indicator phenol red only showed little fluorescence in contact with green light. Even scattering and absorption were negligible.

To test the suitability of both laser types, a collagen-elastin scaffold containing CPOx beads and mesenchymal stem cells was prepared. The scaffold had 10 mm diameter and approximately 14 mm height. The CPOx beads and the cells were located in the center of the scaffold. Pre-breading in a petri dish was performed. Afterwards the scaffold was transferred into the laminar flow bioreactor. The preparation procedure has been previously described in detail in [40]. The oxygen concentration within the scaffold is measured from afar with the modified OPAL system. The measurement interval was one day. Every oxygen measurement was performed twice. Once using the Nd:YAG solid state laser and again after changing the setup to the diode laser as excitation source. Figure 16 shows the results.

Within the error limits, the oxygen concentrations measured with both lasers were identical. From a technical point of view both excitation light sources were equally suitable. The curves showed an increase of the oxygen concentration inside the scaffold. In comparison to petri dish, a better oxygen supply due to the flow of the medium through the scaffold inside the reactor was expected. However, a

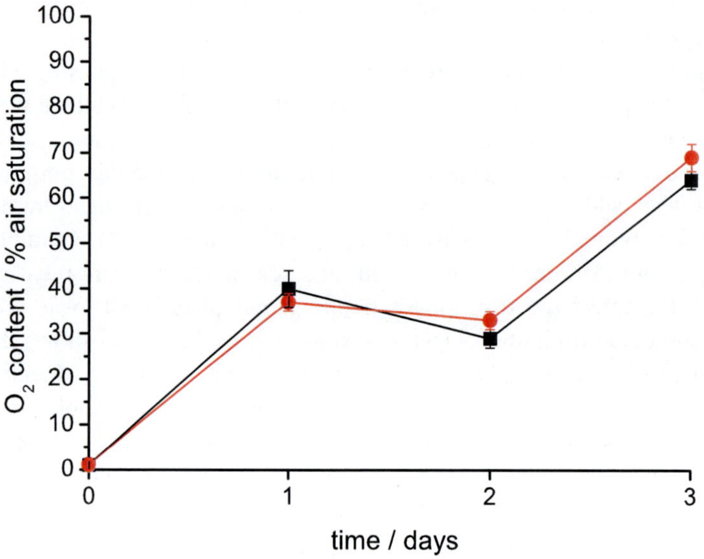

Fig. 16 Optical oxygen measurements within the cell carrier in the center of the laminar flow bioreactor achieved with an optically referenced 532 nm frequency doubled solid-state laser (red circles) and a 515 nm diode laser (black squares)

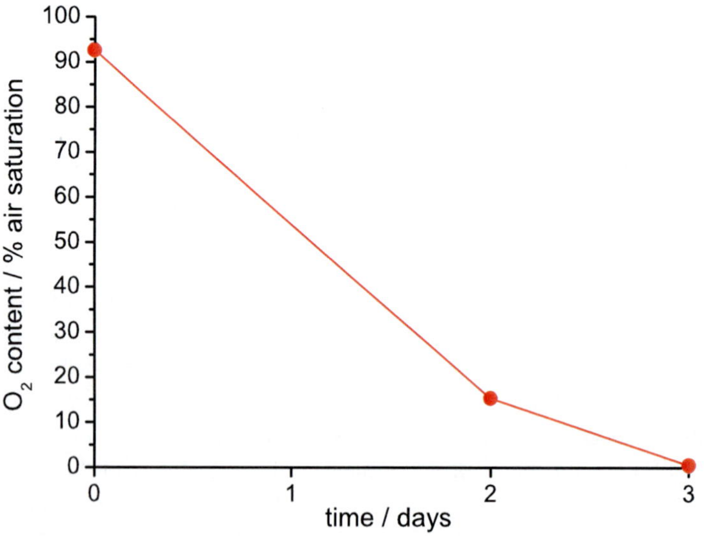

Fig. 17 Complete decrease of the oxygen content in the laminar flow bioreactor due to a fungal infection

sustaining increase indicated a declining cell activity or even cell death. Obviously the culture was affected during the transfer process from the petri dish to the reactor and did not recover within 3 days.

Figure 17 shows another oxygen test series inside a scaffold within the bioreactor archieved with the solid state laser. After 4 days the oxygen content was zero due to a fungal infection that resulted in complete consumption of all oxygen caused by the quickly growing fungal cells.

4 Conclusion

Measuring the phosphorescence decay time of microbeads is a feasible technique to quantify the oxygen content in situ and in real-time in the immediate vicinity of cell tissue. In contrast to traditional electrochemical oxygen probes the measurements can easily performed in closed compartments since no connection wires are necessary. Using a laser as an excitation source, such measurements can also be performed from a distance, e.g. within a macroscopic bioreactor. Interfering fluorescence background, which arises when the laser beam passes the culture medium, is masked by a two frequency phase modulation technique. Phase fluctuations and phase hopping of a solid state laser was eliminated by extending the existing measurement setup with addition of an optical reference channel. The versatility of this technique has been demonstrated for continuous, non-invasive measurement of oxygen concentrations in biological samples in respect to fast changing states, intracellular versus pericellular values and long distances to determine oxygen content within three-dimensional tissue blocks. There are plans to realize a robotic high-throughput screening system in the future.

Acknowledgements The modification of the OPAL oxygen measurement was funded by the investment bank of the German federal state of Brandenburg, grant no. 80149436 (FeLas3D). This work was supported by a grant of Hannover Impuls Excellence Initiative, a State doctorate grant for female scientists for B.W. through Hannover Medical School and a medical student doctoral grant through the "Struc Med" program of Hannover Biomedical Research School for M.N.

References

1. Parmar, K., Mauch, P., Vergilio, J.-A., Sackstein, R., Julian, D.D.: Distribution of hematipoietic stem cells in the bone marrow according to regional hypoxia. PNAS **104** (13), 5431–5436 (2007)
2. Zhang, K., Zhu, L., Fan, M.: Oxygen, a key factor regulating cell behavior during neurogenesis and cerebral diseases: Frontiers in Molecular Neuroscience **4**(5) 1–11 (2011)
3. Garcia-Ochoa, F., Gomez, E.: Bioreactor scale-up and oxygen transfer rate in microbial processes: An overview. Biotechnol. Adv. **27**(2), 153–176 (2009)

4. Li, S., Oreffo, R.O.C., Sengers, B.G., Tare, R.S.: The effect of oxygen tension on human articular chondrocyte matrix synthesis: integration of experimental and computational approaches: Biotechnol. Bioeng. **111**(9), 1876–1885 (2014)
5. Suresh, S., Srivastava, V.C., Mishra, I.M.: Techniques for oxygen transfer measurement in bioreactors: a review. J. Chem. Technol. Biotechnol. **84**, 1091–1103 (2009)
6. Clark, L.C., Wolf, R., Granger, D., Taylor, Z.: Continuous recording of blood oxygen tensions by polarography. J. Appl. Physiol. **6**, 189–193 (1953)
7. Stamati, K., Mudera, V., Cheema, U.: Evolution of oxygen utilization in multicellular organisms and implications for cell signalling in tissue engineering: J. Tissue Eng. **2**(1) (2011). https://doi.org/10.1177/2041731411432365
8. Krawczyk-Bärsch, E., Grossmann, K., Arnold, T., Hofmann, S., Wobus, A.: Influence of uranium (VI) on the metabolic activity of stable multispecies biofilms studied by oxygen microsensors and fluorescence microscopy. Geochim. Cosmochim. Acta **72**(21), 5251–5265 (2008)
9. Volkmer, E., Drosse, I., Otto, S., Stangelmayer, A., Stengele, M., Kallukalam, B.C., Mutschler, W., Schieker, M.: Hypoxia in static and dynamic 3D culture systems for tissue engineering of bone. Tissue Eng A. **14**(8), 1331–1340 (2008)
10. Klimant, I., Ruckruh, F., Liebsch, G., Stangelmayer, A., Wolfbeis, O.S.: Fast Response Oxygen Micro-Optodes Based on Novel Soluble Ormosil Glasses. Mikrochim. Acta **131**, 35–46 (1999)
11. Koo Lee, Y.-E., Kopelman, R.: Chapter twenty-one–Nanoparticle PEBBLE Sensors in Live Cells. Methods in Enzymology. **504** 419–470 (2012)
12. Fercher, A., Borisov, S.M., Zhdanov, A.V., Klimant, I., Papkovsky, D.B.: Intracellular O2 sensing probe based on cell-penetrating phosphorescent nanoparticles. ACS Nano **5**(7), 5499–5508 (2011)
13. Lee Koo, Y.-E., Cao, Y., Kopelman, R., Man Koo, S., Brasuel, M., Philbert, M.-A.: Real-Time Measurements of Dissolved Oxygen Inside Live Cells by Organically Modified Silicate Fluorescent Nanosensors. Anal. Chem. **76**, 2498–2505 (2004)
14. Monson, E., Brasuel, M., Philbert, M.A., Kopelman, R.: PEBBLE nanosensors for in vitro bioanalysis. Tuan Vo-Dinh, Biomedical Photonics Handbook, ISBN0849311160, CRC Press, March 26 (2003)
15. Dmitriev, R.I., Papkovsky, D.B.: Optical probes and techniques for O_2 measurement in live cells and tissue. Cell. Mol. Life Sci. **69**, 2025–2039 (2012)
16. Stepinac, T.K., Chamot, S.R., Rungger-Brändle, E., Ferrez, P., Munoz, J.-L., van den Bergh, H., Riva, C.E., Pournaras, C.J., Wagnières, G.A.: Light-induced retinal vascular damage by Pd-porphyrin luminescent oxygen probes. IOVS. **46**(3), 956–966 (2005)
17. Acosta, M.A., Ymele-Leki, P., Kostov, Y.V., Leach, J.B.: Fluorescent microparticles for sensing cell microenvironment oxygen levels within 3D scaffolds. Biomaterials **30**, 3068–3074 (2009)
18. Wang, X., Otto, S., Wolfbeis, O.S.: Optical methods for sensing and imaging oxygen: materials, spectroscopies and applications. Chem. Soc. Rev. **43**(10), 3666–3761 (2014)
19. Amao, Y.: Probes and Polymers for Optical Sensing of Oxygen. Microchim. Acta **143**, 1–12 (2003)
20. Ast, C., Schmälzlin, E., Löhmannsröben, H.-G., van Dongen, J.T.: Optical Oxygen Micro- and Nanosensors for Plant Applications. Sensors. **12**(6), 7015–7032 (2012)
21. Enko, B., Borisov, S.M., Regensburger, J., Bäumler, W., Gescheidt, G., Klimant, I.: Singlet Oxygen-Induced Photodegradation of the Polymers and Dyes in Optical Sensing Materials and the Effect of Stabilizers on These Processes. J. Phys. Chem. A **117**, 8873–8882 (2013)
22. Benson, B.B., Krause, D.: The concentration and isotopic fractionation of gases dissolved in fresh-water in equilibrium with the atmosphere. 1. Oxygen. Limnol. Oceanogr. **26** 662–671 (1980)
23. Bossi, M.L., Daraio, M.E., Aramendía, P.F.: Luminescence quenching of Ru(II) complexes in polydimethylsiloxane sensors for oxygen. J. Photochem. Photobiol. A: Chemistry **120**, 15–21 (1999)

24. Cai, Y., Smith, A., Shinar, J., Shinar, R.: Data analysis and aging in phosphorescent oxygen-based sensors. Sensors and Actuators B: Chemical. **146**(1), 14–22 (2010)
25. Woods, R.J., Scypinski, S., Cline Love, L.J.: Transient Digitizer for the Determination of Microsecond Luminescence Lifetimes. Anal. Chem. **56**, 1395–1400 (1984)
26. Lakowicz, J.R.: Principles of Fluorescence Spectroscopy, 2nd edn. Kluwer Academic/Plenum Publishers, New York (1999)
27. Schmälzlin, E., van Dongen, J.T., Klimant, I., Marmodée, B., Steup, M., Fisahn, J., Geigenberger, P., Löhmannsröben, H.-G.: An optical multifrequency phase-modulation method using microbeads for measuring intracellular oxygen concentrations in plants. Biophys. J. **89**(2), 1339–1345 (2005)
28. Schmälzlin, E., Friedmann, L., Horn, E., Zantl, R.: Spatially-resolved oxygen measurements in biological samples. Laser+Photonics, 71–73 (2015)
29. Ikai, M., Ishikawa, F., Aratani, N., Osuka, A., Kawabata, S., Kajioka, T., Takeuchi, H., Fujikawa, H., Taga, Y.: Enhancement of External Quantum Efficiency of Red Phosphorescent Organic Light-Emitting Devices with Facially Encumbered and Bulky PtII Porphyrin Complexes. Adv. Funct. Mater. **16**, 515–519 (2006)
30. Shahroosvand, H., Abbasi, P., Mohajerani, E., Janghouri, M.: Red electroluminescence of ruthenium sensitizer functionalized by sulfonate anchoring groups. Dalton Trans. **43**, 9202–9215 (2014)
31. Lehner, P., Staudinger, P., Borisov, S.M., Regensburger, Klimant, I.: Intrinsic Artefacts in Optical Oxygen Sensors—How Reliable are our Measurements? Chem. Eur. J. **21**, 3978 – 3986 (2015)
32. Prill, E., Bavli, D., Levy, G., Ezra, E., Schmälzlin, E., Jaeger, M.S., Schwarz, M., Duschl, C., Cohen, M., Nahmias, Y.: Real-time monitoring of oxygen uptake in hepatic bioreactor shows CYP450-independent mitochondrial toxicity of acetaminophen and amiodarone. Arch. Toxicol. **90**, 1181–1191 (2016)
33. Bavli, D., Prill, S., Ezra, E., Levy, G., Cohen, M., Vinken, M., Vanfleteren, J., Jaeger, M., Nahmias, Y.: Real-time monitoring of metabolic function in liver-onchip microdevices tracks the dynamics of mitochondrial dysfunction. PNAS **113**(16), E2231–E2240 (2016)
34. Prill, S., Anderson, A., Papkovsky D.B., Schmälzlin, E.: Intracellular O_2 Measurements. G.I. T. Laboratory Journal Europe. **1–2**, 28–29 (2014)
35. Ethirajan, M., Chen, Y., Joshi, P., Pandey, R.K.: The role of porphyrin chemistry in tumor imaging and photodynamic therapy. Chem. Soc. Rev. **40**, 340–362 (2011)
36. Israelowitz, M., Weyand, B., Rizvi, S., Vogt, P.M., von Schroeder, H.P.: Development of a Laminar Flow Bioreactor by Computational Fluid Dynamics. J. Healthcare Engineering **3**(3), 455–476 (2012)
37. Grzelak, A., Rychlik, B., Bartosz, G.: Light-dependent generation of reactive oxygen species in cell culture media. Free Radic. Biol. Med. **30**(12), 1418–1425 (2001)
38. Cardoso, D.R., Libardia, S.H., Skibsted L.H.: Riboflavin as a photosensitizer. Effects on human health and food quality. Food Funct. **3** 487–502 (2012)
39. Wilkinson, F., Helman, W.P., Ross, A.B.: Rate Constants for the Decay and Reactions of the Lowest Electronically Excited Singlet State of Molecular Oxygen in Solution. An Expanded and Revised Compilation. Journal of Physical and Chemical Reference Data. **24**(2) 663–1021 (1995)
40. Weyand, B., Nöhre, M., Schmälzlin, E., Stolz, M., Israelowitz, M., Gille, C. von Schroeder, H.P., Reimers, Vogt, P.M.: Noninvasive Oxygen Monitoring in Three-Dimensional Tissue Cultures Under Static and Dynamic Culture Conditions. BioResearch Open Access **4**(1) 266–277 (2015)

Emerging Biomimetic Approaches in the Optimization of Drug Therapies

Obaro S. Michael

Abstract Pharmacology is the study of drug action and applications in living organisms. A significant aspect of pharmacology is the use of natural, semi-synthetic, and synthetic products in the prophylaxis or treatment of disease. Observations from nature are increasingly being adopted and adapted to solve real world challenges. Potential solutions to challenges of restoring biological systems to normal function may lie in normally functioning biological systems. Biomimetics is learning from nature and using nature's methods in science and engineering. This review is based on a literature search of Medline conducted within the first week of October 2016, limited to articles published between 2005 and 5th October 2016. Search terms were Biomimetics and Pharmacology. The most recent 100 publications were used to determine emerging themes and interphases of Biomimetics and Pharmacology. Using the 100 most recent publications from the search, the emerging themes were biomaterial scaffolds (26%), bio-molecules with enhanced or novel pharmacologic properties (24%), nano-particles (11%), cancer therapeutics (9%), biomimetics method applied to pharmacology (6%), novel and improved antimicrobials (6%), stem cell science (4%), regenerative medicine (3%), siRNA molecules (2%), and molecule biosensors (1%). This review describes these themes using a few examples from literature. Advances in biomimetic research are rapidly expanding the scope of pharmacology and creating new themes and trends. These emerging themes are giving rise to new sub-specialties in pharmacology. However, great science will remain a multidisciplinary collaboration in one big venture—to keep improving the quality of life on earth.

Keywords Biomimetics · Pharmacology · Biomimetics—Pharmacology

O. S. Michael (✉)
Chemotherapy Unit, Department of Pharmacology and Therapeutics College of Medicine, University of Ibadan, Ibadan, Nigeria
e-mail: micobaro@gmail.com

M. Israelowitz et al. (eds.), *Biomimetics and Bionic Applications with Clinical Applications*, Series in BioEngineering,
https://doi.org/10.1007/978-3-319-53214-1_9

1 Background

1.1 Scope of Pharmacology

Pharmacology involves the study of drug action and applications in living organisms. A significant aspect of pharmacology is the use of natural, semi-synthetic, and synthetic products in the prophylaxis or treatment of disease. Drugs are molecules that have the real or potential capacity to alter biochemical processes in living systems. To achieve this, majority of drugs must be able to cross membrane barriers and bind to macromolecules called receptors. Clinical pharmacology is the study of drug use in the treatment of human diseases. The ultimate goal of pharmacology is having drugs with minimal or no side effects, no adverse drugs reactions, but with maximal efficacies [1]. This has remained a challenge that pharmacologists and biological scientists have continued to be engaged in. Clinical applications remain the driving force behind advancements in pharmacology.

In its early days, before the advent of synthetic organic chemicals with potent medicinal activities, pharmacology concerned itself with understanding the effects of natural substances, mainly plant extracts. Friedrick Sertuner (Fig. 1), a young German pharmacist, reported the purification of morphine from opium in 1805 [2]. To have an idea of the effects of the drug, Sertürner was able to confirm the hypnotic as well as the analgesic properties of morphine using animal models.

Fig. 1 Friedrich Wilhelm Adam Serturner (19 June 1783–20 February 1841). German pharmacist, who discovered morphine in 1804. Downloaded from the Alchetron website http://alchetron.com/Friedrich-Serturner-1111583-W on Wednesday 12 October 2016 (with permission)

Other pharmacologists followed, with the identification of many plant products with medicinal properties such as quinine, digitalis, atropine, ephedrine, strychnine, and many others—majority of which are still in use today. The majority of new drugs have been generated from crude extracts natural products, secondary metabolites and refined compounds derived from natural products [3–6]. While most reports from the early days of pharmacology focussed on the effects on drugs on living systems and models of disease, recent advances in pharmacology have focussed more on the molecular mechanism of drug action. Relatively smaller aspects of these advances have begun to examine biomimetics and applications in optimization of drug delivery and efficacy.

1.2 Biomimetics in Pharmacology

Observations from nature are increasingly being adopted and adapted to solve real world challenges [7]. Potential solutions to challenges of restoring biological systems to normal function may lie in normally functioning biological systems. Biomimetics, in a sense, is learning from nature and using nature's methods in science and engineering. Developments in bioengineering are beginning to show that mimicking nature may be the better option in solving real life challenges, enhancing human abilities, and restoring function where disabilities exist.

Pharmacology involves the use of molecules and substances that affect, enhance, or alter functions in living systems. Over billions of years, nature has developed elaborate living systems that have colonized almost every space on earth. Life forms have established successful colonies in water, on surfaces of seas, on land, in deserts, thick forests, and in the air. All manner of relationships exist between microorganisms and their different hosts. These are avenues for unique solutions to challenges of environmental and biological health. From nature, there are limitless variations to approach of utilization and conservation of energy sources.

Drug delivery remains a challenging aspect of pharmacology. Barriers to effective drug delivery include physic-chemical barriers like pH and anatomical barriers like the Blood-Brain Barrier [7]. In addition, it has been difficult to develop drugs with minimal side effects; occasionally these side effects may be fatal in certain individuals [8]. These challenges need to be solved and today reports are showing that biomimetics may offer exceptional solutions to some of the most difficult challenges of pharmacology. In this chapter, some emerging applications of biomimetics in some selected difficult areas of pharmacology are described. The pharmacologist still pursues the 'magic bullet'—a drug with maximized efficacy and no side effects—biomimetics may bring us closer to, or even into, this dream.

2 Methods

This review is based on a literature search of Medline (PUBMED) conducted within the first week of October 2016, limited to articles published between 2005 and 5th October 2016. Search terms were Biomimetics and Pharmacology. The NCBI search details were *(("biomimetics"[MeSH Terms] OR "biomimetics"[All Fields]) AND ("pharmacology"[Subheading] OR "pharmacology"[All Fields] OR "pharmacology"[MeSH Terms])) AND ("2005"[PDAT]: "3000"[PDAT])*. Retrieved manuscripts were in the field of the interphase between biomimetics and pharmacology. Examination of the backgrounds, methodologies, results, and conclusions of the retrieved manuscripts were done to determine the nature of interphases between biomimetics and pharmacology. This was used to determine the emerging themes. Emerging theme in this review refers to specialized areas of research in biomimetics with direct relevance and linkage to pharmacology. The most recent 100 Manuscripts returned by the search were entered onto Statistical Software for the Social Sciences (SPSS) version 20 and coded in line with their specialized areas of research (Biomimetics-Pharmacology). Similar areas of research themes were grouped together and relative frequencies of these themes were determined. The discussion of these themes was done using a few publications selected from the search or other more illuminating publications.

2.1 Biomimetics—Pharmacology Themes

The themes that were most common in applications of Biomimetics to pharmacology were obtained from the examined literature and described. This review is limited, and not exhaustive, as not all retrieved references are discussed. However, the arguments are representative of the rising importance of Biomimetics in the field of pharmacology particularly in several aspects of drug discovery, design, and delivery.

3 Results

3.1 Biomimetics and Pharmacology Themes

Using the 100 most recent publications from the search as described in the methodology section, the themes and frequency of publications were biomaterial scaffolds (26%), bio-molecules with enhanced or novel pharmacologic properties (24%), nano-particles (11%), cancer therapeutics (9%), biomimetics method applied to pharmacology (6%), novel and improved antimicrobials (6%), stem cell science (4%), regenerative medicine (3%), siRNA molecules (2%), and molecule biosensors (1%).

3.2 Biomaterial Scaffolds

Biomimetic scaffolds are increasingly being designed and developed worldwide. Cellulose-based scaffolds have been produced by electrospinning of cellulose acetate solution followed by saponification with NaOH/ethanol and can be fashioned into 3D cellulose sponges [9]. The 3D cellulose sponge has good bone tissue engineering applications; mimicking natural 3D designs have shown better performances compared to 2D cellulose mats. Scaffolds have many uses that expand into pharmacologic applications. A significant proportion of drug receptors are membrane bound and are proteins. Cell membranes have a bi-lipid structure to which proteins and natural macromolecules are attached. The myriad functions of proteins are difficult to evaluate in artificial systems. Using biomimetics, bi-lipid scaffolds have been designed to study the natural biochemical pathways in living systems. Biomimetic approaches based on supported lipid bilayers have been used to explore the interactions between complement proteins and inhibitors, thereby offering insights into membrane attack complex (MAC) assembly and regulation [10].

Drug action can be significantly enhanced by improvements in drug delivery methods. A method of maintaining constant drug levels in living system is by developing drugs with altered physicochemical properties that result in slowed absorption in living systems. Biomimetic scaffolds are providing avenues for controlled release of drugs. Biomimetic supra-molecular designs have been applied to controlled release of growth factors in bone regeneration with promising results [11]. Supra-molecular designs have potential uses in delivery of antibiotics, anti-hypertensives, and other drugs. Scaffolds designed to have potent antimicrobial activities that may be used in wound dressings and in coating tubes used in invasive procedures [12–15]. Recent advances in the design and preparation of supra-molecular delivery systems have clear potentials for translation to clinical uses.

3.3 Bio-Molecules with Enhanced Pharmacologic Properties

Central to pharmacology are molecules with activity in living systems. Fundamentally, drug action is drug molecules affecting receptor sites. The specificity of drug action often depends on selectivity and receptor affinity. Designing drugs with minimal side effects is an ongoing process in drug discovery, design, and production. Biomimetics has opened up new vistas of developing drugs with better spectrums of activity.

Numerous examples of applications of biomimetics to molecular design for enhanced activity or reduced adverse effects exist. Peptides are biologically occurring short chains of amino acid monomers linked by amide bonds. Peptides have potential uses in therapeutics but the major hindrance to peptide drug development remains their short half-lives in vivo due to enzymatic degradation.

Biomimetic approaches are providing strategies to circumventing this limitation. A strategy for enhancing the in vivo half-life of peptides without compromising their potency involves endowing peptides with a small molecule that binds reversibly to the serum protein transthyretin [16]. It was shown that this strategy was effective in enhancing the half-life of an agonist for Gonadotropin-releasing hormone (GnRH) receptor while maintaining its binding affinity, which was translated into superior in vivo efficacy [16].

Synthesizing peptide mimics is an approach that has been used to develop synthetic molecules with natural peptide activity. A synthetic peptide named MAP27, that mimics Peptidoglycan (PGN), a conserved and major component of *S. aureus* cell wall, may lead to the design of effective vaccines against the bacterium [17]. Designing peptide-mimicking molecules may also be used to design vaccines against many pathogenic agents. Clinically relevant molecules like erythropoietin and heparin have been designed using biomimetic approaches with better profiles of activity [18, 19]. Biomimetic heparin and other molecules with potent pharmacologic activities are increasingly being explored. Molecular bioengineering, guided by biomimetics, will greatly transform the scope and horizons of pharmacology.

3.4 Nanoparticles

Nano-particles have transformed the methods of drug delivery. Recent advances in nanoparticle have led to design of targeted drug delivery systems, better therapeutics, and fewer side effects. The rapid developments in multifunctional nanoparticles provide ample opportunities to integrate both diagnostic and therapeutic modalities into a single effective cancer "theranostic" vector [20]. Nanoparticles could extravasate passively into the tumour tissues in preference to the adjacent normal tissues by capitalizing on the enhanced permeability and retention effect [20]. Tumour targeting might be further augmented by conjugating tumour-specific peptides and antibodies onto the surface of these nanoparticles making them useful methods of better anti-tumour therapeutics and imaging. By targeted imaging using impregnated Nano-molecules and antibody-antigen complexes, cells and smaller structures such as bacteria and viruses may be imaged. Ultrasound-triggered drug delivery is now becoming a mature technology with first patients enrolling in clinical trials. [21]

In tropical countries, infectious diseases remain the major cause of morbidity and mortality. The diseases of poverty such as malaria, tuberculosis, and HIV, remain widespread with limited resources for their control and management. Malaria, a diseases caused by five *Plasmodium* species in humans, remains a major cause death in children and pregnant women in sub-Saharan Africa. The malaria parasite has developed resistance to all known therapeutic agents, resulting in a frantic search for newer and more efficacious antimalarial therapies [22–25]. Nano-particles are offering new hope for better design of antimalarial drugs and more effective delivery systems. The adaptation of existing antimalarial

Nano-carriers to new *Plasmodium* stages, drugs, targeting molecules, or encapsulating structures is a strategy that can provide new nanotechnology-based, cost-efficient therapies against malaria [26].

3.5 Cancer Therapeutics

Drug treatment of tumour cells remains one of the most challenging areas of pharmacology. Tumour cells are almost exactly like normal cells and thus it is very difficult to apply selective toxicity in eliminating them. Using new approaches, Biomimeticians have begun to examine potentially better options at tackling tumour cells. Directly targeting the tumour cells by homing mechanism driven drug delivery has the potential of significantly improving cancer therapy. A new class of therapeutic drugs suitable for the task has emerged based on the concept of virus-mimetic Nano carriers, or 'artificial viruses' [27]. This approach reduces exposure to toxic dose of anticancer regimen.

Another biomimetic approach to improved diagnosis and treatment of cancer is Tumour-Specific Formation of Enzyme-Instructed Supra-molecular Self-Assemblies as Cancer Theranostics [28]. The phosphatase-instructed co-assembly process, as well as its therapeutic and diagnostic capability, has been successfully evaluated at different levels ranging from in vitro, living cell, tissue mimic, to in vivo. Clear differentiation of tumour and normal cells has been achieved using this method. Complete tumour elimination with high therapeutic accuracy has also been successfully achieved upon laser irradiation within 24-48 h post injection. In vivo formation of tumour-specific indocyanine green (ICG) - doped nano-fiber for photo thermal therapy (PTT) theranostics has the immense potential for clinical translation of personalized nanomedicine with targeted drug delivery as well as for cancer diagnosis and therapy [28]. Also being tried are second mitochondrial activator (smac) mimetics that interrupt specific caspase regulators and cellular transport, macrophage mediated biomimetic delivery system for the treatment of lung metastasis of breast cancer, and nanotube interactions with microtubules [29–31]. Interaction between nanotubes and microtubules inside live cells leads to microtubule dysfunction, mitotic arrest and cell death [29]. These exciting approaches could play pivotal roles in future cancer treatments.

3.6 Biomimetic Methods Applications

Methods in biomimetics will transform experimental pharmacology. Only two of these methods will be mentioned here. Red cells are important in physiological wellbeing of primates, including humans. Parasitic diseases like malaria are pathogens of red cells and often result in different grades of anaemia and other morbidities. Effective methods of isolating red cells, cleaning, and restoring them without affecting their functional capacities will be beneficial in experimental pharmacology, diagnosis, and therapeutics of red blood cell conditions. The development of micro-plates for high throughput filtration of RBC through microsphere layers (micro-plate-based microsphiltration) has been undertaken [32]. This process mimic's the clearance of deformed red blood cells by the spleen. This method will impact significantly on research involving isolation of red cells. Potential applications include malaria chemotherapy and blood transfusion.

A signification proportion of pre-clinical trials depend on the use of natural models of disease; cell culture, isolated organ, or whole animal models. These natural models have limitations that include ethical considerations, difficulties with developing appropriate controls, lack of gradual modifications of the models, and wide variations in constitutive compositions. Bioengineered cell cultures may overcome some of these barriers. A new generation of bioengineered tumours is now emerging in response to these limitations, with potential to transform drug screening by providing predictive models of tumours within their tissue context, for studies of drug safety and efficacy [33]. Biomimetic bioengineering has almost limitless applications, a lot of which will provide new models for research in pharmacology.

3.7 Biomimetic Antimicrobials

Pathogenic microbes are microorganisms that cause disease in animals and plants. An antimicrobial is an agent that kills microorganisms or inhibits their growth. Biomimetic methods of antimicrobial drug discovery and development are increasingly being developed, many of which are innovative approaches to drug discovery and development. Zinc oxide is a versatile biocompatible substance used in many applications. As a semiconductor, ZnO has several properties including good transparency, high electron mobility, wide band gap, and strong room-temperature luminescence [34]. These properties are valuable in emerging applications for: transparent electrodes in liquid crystal displays, energy-saving or heat-protecting windows, and electronics as thin-film transistors and light-emitting diodes [35]. More recently, the antibacterial properties of ZnO nanorods are being investigated in both Gram-positive and Gram-negative microorganisms. Overall, experimental results suggest that ZnO nanorods could be developed as antibacterial agents against a wide range of microorganisms to control and prevent the spreading and persistence of bacterial infections [36]. This introduces a new concept to synthesize ZnO nanorods by using egg white as a biological template for various applications including food science, animal science, biochemistry, microbiology and medicine [36]. Similar innovative methods have continued to confirm that biomimetic antimicrobial drug discovery is a novel approach that will revolutionize the field of antimicrobials and their applications.

3.8 Stem Cell Research

Stem cells are undifferentiated biological cells that can differentiate into specialized cells and can divide to produce more stem cells. Stem cell research is one of the most exciting advances in the biological sciences; however, a lot of ethical issues need to be resolved. More common is the use of autologous adult stem cells that can be derived from blood, adipose tissue, and bone marrow. The medical applications of adult stem cells are on the increase worldwide.

Stem cell applications span through all tissues, organs, and diseases in the body. Stem cells have arisen as powerful tools to improve skin wound healing, due to features such as effective secretome, self-renewal, low immunogenicity, and differentiation capacity [37]. They represent potentially readily available biological material that can particularly target distinct wound-healing phases [37]. In this context, mesenchymal stem cells have been shown to promote cell migration, angiogenesis, and a possible regenerative rather than fibrotic microenvironment at the wound site, mainly through paracrine signalling with the surrounding cells/tissues [37].

Relatively recently, stem cells can now be artificially grown and transformed (differentiated) into specialized cell types with characteristics consistent with cells of various tissues such as muscles or nerves. There is the real possibility of discovering a family of molecules (designed and developed into medications) that can specifically reprogram adult cells into stem cells and redirect cellular processes into regenerating specialized cells at will. This potential is still under intense research. Embryonic cell lines and autologous embryonic stem cells generated through somatic cell nuclear transfer or dedifferentiation have been proposed as promising candidates for a host of future therapies [38]. With the concept of cell and tissue therapy, many chronic disorders will be curable [39]. Advances in stem cell research are the corner stone of regenerative medicine.

3.9 Regenerative Medicine

Regenerative medicine may be considered the ultimate goal of research in biomimetics. Regenerative Medicine is an emerging interdisciplinary field of research and clinical applications focused on the repair, replacement or regeneration of cells, tissues or organs to restore impaired function resulting from any cause, including congenital defects, disease, trauma and aging [40]. It uses a combination of several technological approaches that moves it beyond traditional transplantation and replacement therapies. These approaches may include, but are not limited to, the use of soluble molecules, gene therapy, stem cell transplantation, tissue engineering and the reprogramming of cell and tissue types [40]. There has been a steady increase in research in regenerative medicine. Regenerative medicine holds the promise of ending shortages of organs required for transplant, reversing sequelae of cardiac and cerebral infarcts, curing diabetes mellitus, reversing age associated degenerative diseases, and an almost endless list of other pathologies [41, 42].

3.10 siRNA Molecules

Small interfering RNA (siRNA) is a class of double-stranded RNA molecules, 20–25 base pairs in length that interferes with the expression of specific genes with complementary nucleotide sequences by degrading mRNA after transcription resulting in no translation [43]. Given the ability to knock down any gene of interest, siRNA has generated a great deal of interest in both basic and applied biology and pharmacology. Biomimetic molecules designed to function by this process are undergoing pre-clinical trials. They will serve as precisely targeted therapies in many disease conditions. siRNAs are useful in the assessment of the contributions of genes to cellular phenotypes including cytokinesis, apoptosis, insulin signalling, and cell differentiation [44–48]. siRNA screens have been used to identify novel biochemical pathways and have had significant impact in validating targets for a number of cellular processes and diseases including cancer, HIV infection and hepatitis [49–52]. In vivo siRNA has been used for target validation studies in animal disease models with the potential of being used for therapeutic purposes where disease-causing genes are selectively targeted and suppressed [50, 53].

3.11 Molecule Biosensors

A biosensor is an analytical device, used for the detection of an analyte that combines a biological component with a physicochemical detector. Biosensors utilize biomimetic processes to detect the analyte, generate amplifiable signals, which are further processed by a transducer. The physicochemical processes may be any component of the electromagnetic spectrum or a biochemical process that results in the generation of a signal. Applications of biosensor systems are very diverse and universal, especially in biological disciplines. These include glucose monitoring, receptor-ligand interactions, metabolites detectors, all modalities of imaging, trace gas/element detectors, electronic thermometers, digital meters, micro arrays, nanoscale biosensors, etc.

Biomimetic methods are providing better biosensor designs for accurate differentiation of molecules in living systems. A biomimetic enzyme modified electrode for H_2O_2 highly sensitive detection has been developed and tried with good results [54]. Another publication reports the construction of a photoelectrochemical biosensing chip using mussel-inspired polydopamine coating strategy, which demonstrated improved photo-to-electric conversion performance for the CdS/

TiO2-ITO chip, and was used for the direct immobilization of captured antibodies and the detection of CD146 in the absence of additional electron donors/acceptors [55]. A fast, cheap and easy to use analytical system for detecting dioxins was designed using a biomimetic approach [56]. Automated Biomimetic systems for accurate drug level assays are still ahead in the future, but they will transform current methods of clinical pharmacokinetics.

4 Conclusions

Biomimetic research is still at an early phase of development but the potential impact on pharmacology is obvious. Over 90% of the recent advances described in this review, and those being reported, are still at the phase of pre-clinical trials. Hopes are high that these novel developments will pass pre-clinical and clinical trial phases and become available as therapeutics options for the amelioration and cure of human and animal diseases. Although much less research is being done on plants, the potential of improving crop yields in agriculture are being explored [57]. Malnutrition remains a major challenge in many developing countries, and exploring applications of Biomimetic research in improving agricultural yield and quality will become a priority area globally.

From a pharmacologist point of view, the field of pharmacology is being rapidly transformed. These novel advances are expanding the scope of the subject. Pharmacology in the future will be more of integrative interdisciplinary thematic departments and units rather than the traditional general and systemic ones of today. For example, Nano-particles have led to the new field of Nano-pharmacology. Nano-pharmacology is a branch of pharmacology which investigates interaction of a nanomedicine with living systems at the nanoscale level [58]. Nanoparticles will revolutionize methods of drug delivery.

Stem cell research and regenerative medicine have given birth to regenerative pharmacology. Regenerative pharmacology has been defined as the application of pharmacological sciences to accelerate, optimize, and characterize (either in vitro or in vivo) the development, maturation, and function of bioengineered and regenerating tissues [42]. As such, regenerative pharmacology seeks to cure disease through restoration of tissue/organ function [42]. This strategy is distinct from standard pharmacotherapy, which is often limited to the amelioration of symptoms [42]. All the emerging themes may give rise to new sub-specialties in pharmacology. However, great science will remain a multidisciplinary collaboration in one big venture—to keep improving the quality of life on earth.

Acknowledgements I give credit to Dharmacon, RNAi, Gene expression and Gene editing (http://dharmacon.gelifesciences.com/applications/rna-interference/sirna/) for material on the segment on siRNA. I acknowledge the support of Dr. Meir Israelowitz, Dr. Birgit Weyand, Prof Kerstin Reimers (of blessed memory), Prof Peter Vogt and colleagues at Hanover Medical School, Germany, for supporting my wife and me through the years and showing us developments in Biomimetics and plastic surgery. I acknowledge Dr Afieharo Michael (my wife) for her contributions to this chapter and to my career. I acknowledge my mentor Professor Catherine Falade for her support and encouragement through the years. Credit goes to many more people and sources that I wish to say all are duly recognized with gratitude. I thank God for Grace, for all the support, and encouragement given to me towards completing this work.

Author's Declaration The author declares no competing interests.

References

1. de Chaisemartin, L., Loriot, M.A.: Pharmacogenetics of anticancer drugs. Pathol. Biol. (Paris) **53**, 116–124 (2005)
2. Goerig, M., Esch, J.: Friedrich Wilhelm Adam Sertürner—the discoverer of morphine Anasthesiol Intensivmed Notfallmed Schmerzther **26**, 492–498 (1991)
3. Chang, J., Kwon, H.J.: Discovery of novel drug targets and their functions using phenotypic screening of natural products. J. Ind. Microbiol. Biotechnol. **43**, 221–231 (2016)
4. Williams, L.A., O'Connar, A., Latore, L., Dennis, O., Ringer, S., Whittaker, J.A., Conrad, J., Vogler, B., Rosner, H., Kraus, W.: The in vitro anti-denaturation effects induced by natural products and non-steroidal compounds in heat treated (immunogenic) bovine serum albumin is proposed as a screening assay for the detection of anti-inflammatory compounds, without the use of animals, in the early stages of the drug discovery process. West Indian Med. J. **57**, 327–331 (2008)
5. Luesch, H., Abreu, P.: A natural products approach to drug discovery: probing modes of action of antitumor agents by genome-scale cDNA library screening. Methods Mol. Biol. **572**, 261–277 (2009)
6. Lahlou, M.: Screening of natural products for drug discovery. Expert Opin. Drug Discov. **2**, 697–705 (2007)
7. Hwang, J., Jeong, Y., Park, J.M., Lee, K.H., Hong, J.W., Choi, J.: Biomimetics: forecasting the future of science, engineering, and medicine. Int. J. Nanomed. **10**, 5701–5713 (2015)
8. Ganigara, A., Ravishankar, C., Ramavakoda, C., Nishtala, M.: Fatal hyperkalemia following succinylcholine administration in a child on oral propranolol. Drug. Metab. Pers. Ther. **30**, 69–71 (2015)
9. Joshi, M.K., Pant, H.R., Tiwari, A.P., Maharjan, B., Liao, N., Kim, H.J., Park, C.H., Kim, C. S.: Three-dimensional cellulose sponge: fabrication, characterization, biomimetic mineralization, and in vitro cell infiltration. Carbohydr. Polym. **136**, 154–162 (2016)
10. Yorulmaz, S., Jackman, J.A., Hunziker, W., Cho, N.J.: Supported lipid bilayer platform to test inhibitors of the membrane attack complex: insights into biomacromolecular assembly and regulation. Biomacromol. **16**, 3594–3602 (2015)
11. Azevedo, H.S., Pashkuleva, I.: Biomimetic supramolecular designs for the controlled release of growth factors in bone regeneration. Adv. Drug Deliv. Rev. **94**, 63–76 (2015)
12. Fisher, L.E., Yang, Y., Yuen, M.F., Zhang, W., Nobbs, A.H., Su, B.: Bactericidal activity of biomimetic diamond nanocone surfaces. Biointerphases **11**(1), 011014. https://doi.org/10.1116/1.4944062. PMID: 26992656.011014 (2016)
13. Archana, D., Dutta, J., Dutta, P.K.: Evaluation of chitosan nano dressing for wound healing: characterization, in vitro and in vivo studies. Int. J. Biol. Macromol. **57**, 193–203 (2013)
14. Kim, J.I., Pant, H.R., Sim, H.J., Lee, K.M., Kim, C.S.: Electrospun propolis/polyurethane composite nanofibers for biomedical applications. Mater. Sci. Eng. C Mater. Biol. Appl. **44**, 52–57 (2014)
15. Mandal, A., Sekar, S., Seeni Meera, K.M., Mukherjee, A., Sastry, T.P., Mandal, A.B.: Fabrication of collagen scaffolds impregnated with sago starch capped silver nanoparticles

suitable for biomedical applications and their physicochemical studies. Phys. Chem. Chem. Phys. **16**, 20175–20183 (2014)
16. Penchala, S.C., Miller, M.R., Pal, A., Dong, J., Madadi, N.R., Xie, J., Joo, H., Tsai, J., Batoon, P., Samoshin, V., Franz, A., Cox, T., Miles, J., Chan, W.K., Park, M.S., Alhamadsheh, M.M.: A biomimetic approach for enhancing the in vivo half-life of peptides. Nat. Chem. Biol. **11**, 793–798 (2015)
17. Wang, X.Y., Huang, Z.X., Chen, Y.G., Lu, X., Zhu, P., Wen, K., Fu, N., Liu, B.Y.: A multiple antigenic peptide mimicking peptidoglycan induced T cell responses to protect mice from systemic infection with staphylococcus aureus. PLoS ONE **10**, e0136888 (2015)
18. Guarnieri, F.: Designing a small molecule erythropoietin mimetic. Methods Mol. Biol. **1289**, 185–210 (2015)
19. Farrugia, B.L., Lord, M.S., Melrose, J., Whitelock, J.M.: Can we produce heparin/heparan sulfate biomimetics using "mother-nature" as the gold standard? Mol. **20**, 4254–4276 (2015)
20. Datta, N.R., Krishnan, S., Speiser, D.E., Neufeld, E., Kuster, N., Bodis, S., Hofmann, H.: Magnetic nanoparticle-induced hyperthermia with appropriate payloads: Paul Ehrlich's "magic (nano)bullet" for cancer theranostics? Cancer Treat. Rev. **50**, 217–227 (2016)
21. Boissenot, T., Bordat, A., Fattal, E., Tsapis, N.: Ultrasound-triggered drug delivery for cancer treatment using drug delivery systems: from theoretical considerations to practical applications. J. Control. Release **241**, 144–163 (2016)
22. Wang, Z., Cabrera, M., Yang, J., Yuan, L., Gupta, B., Liang, X., Kemirembe, K., Shrestha, S., Brashear, A., Li, X., Porcella, S.F., Miao, J., Yang, Z., Su, X.Z., Cui, L.: Genome-wide association analysis identifies genetic loci associated with resistance to multiple antimalarials in plasmodium falciparum from China-Myanmar border. Sci. Rep. **6**, 33891 (2016)
23. Woodrow, C.J., White, N.J.: The clinical impact of artemisinin resistance in Southeast Asia and the potential for future spread. FEMS Microbiol. Rev. (2016)
24. Avitia-Dominguez, C., Sierra-Campos, E., Betancourt-Conde, I., Aguirre-Raudry, M., Vazquez-Raygoza, A., Luevano-De la Cruz, A., Favela-Candia, A., Sarabia-Sanchez, M., Rios-Soto, L., Mendez-Hernandez, E., Cisneros-Martinez, J., Palacio-Gastelum, M. G., Valdez-Solana, M., Hernandez-Rivera, J., De Lira-Sanchez, J., Campos-Almazan, M., Tellez-Valencia, A.: Targeting plasmodium metabolism to improve antimalarial drug design. Curr. Protein Pept. Sci. **17**, 260–274 (2016)
25. Lawal, B., Shittu, O.K., Kabiru, A.Y., Jigam, A.A., Umar, M.B., Berinyuy, E.B., Alozieuwa, B.U.: Potential antimalarials from African natural products: a review. J. Intercult. Ethnopharmacol. **4**, 318–343 (2015)
26. Marques, J., Valle-Delgado, J. J., Urban, P., Baro, E., Prohens, R., Mayor, A., Cistero, P., Delves, M., Sinden, R. E., Grandfils, C., de Paz, J. L., Garcia-Salcedo, J. A. & Fernandez-Busquets, X.: Adaptation of targeted nanocarriers to changing requirements in antimalarial drug delivery. Nanomed (2016)
27. Unzueta, U., Cespedes, M.V., Vazquez, E., Ferrer-Miralles, N., Mangues, R., Villaverde, A.: Towards protein-based viral mimetics for cancer therapies. Trends Biotechnol. **33**, 253–258 (2015)
28. Huang, P., Gao, Y., Lin, J., Hu, H., Liao, H.S., Yan, X., Tang, Y., Jin, A., Song, J., Niu, G., Zhang, G., Horkay, F., Chen, X.: Tumor-specific formation of enzyme-instructed supramolecular self-assemblies as cancer theranostics. ACS Nano **9**, 9517–9527 (2015)
29. Garcia-Hevia, L., Fernandez, F., Gravalos, C., Garcia, A., Villegas, J.C., Fanarraga, M.L.: Nanotube interactions with microtubules: implications for cancer medicine. Nanomed. (Lond) **9**, 1581–1588 (2014)
30. Fu, J., Wang, D., Mei, D., Zhang, H., Wang, Z., He, B., Dai, W., Wang, X., Zhang, Q.: Macrophage mediated biomimetic delivery system for the treatment of lung metastasis of breast cancer. J Control Release **204**, 11–19 (2015)
31. Talbott, R.L., Borzilleri, R.M., Chaudhry, C., Fargnoli, J., Shen, H., Fairchild, C., Barnhart, B., Ortega, M., McDonagh, T.E., Vuppugalla, R., Vite, G.D., Hunt, J.T., Gottardis, M., Naglich, J.G.: Pharmacology of smac mimetics; chemotype differentiation based on physical association with caspase regulators and cellular transport. Exp. Cell Res. **338**, 251–260 (2015)

32. Duez, J., Holleran, J.P., Ndour, P.A., Pionneau, C., Diakite, S., Roussel, C., Dussiot, M., Amireault, P., Avery, V.M., Buffet, P.A.: Mechanical clearance of red blood cells by the human spleen: potential therapeutic applications of a biomimetic RBC filtration method. Transfus. Clin. Biol. **22**, 151–157 (2015)
33. Villasante, A., Vunjak-Novakovic, G.: Tissue-engineered models of human tumors for cancer research. Expert Opin. Drug Discov. **10**, 257–268 (2015)
34. Fan, Z., Lu, J.G.: Zinc oxide nanostructures: synthesis and properties. J. Nanosci. Nanotechnol. **5**, 1561–1573 (2005)
35. Ozgur, U.A.Y., Liu, C., Teke, A., Reshchikov, M., Dogan, S., Avrutin, V., Cho, S., Morkoc, H.: A comprehensive review of ZnO materials and devices. Appl. Phy. Rev. **98**, 1–103 (2005)
36. Ahmed, F., Arshi, N., Jeong, Y.S., Anwar, M.S., Dwivedi, S., Alsharaeh, E., Koo, B.H.: Novel biomimatic synthesis of ZnO nanorods using egg white (albumen) and their antibacterial studies. J. Nanosci. Nanotechnol. **16**, 5959–5965 (2016)
37. Cerqueira, M.T., Pirraco, R.P., Marques, A.P.: Stem cells in skin wound healing: are we there yet? Adv. Wound. Care (New Rochelle) **5**, 164–175 (2016)
38. Tuch, B.E.: Stem cells—a clinical update. Aust. Fam. Phys. **35**, 719–721 (2006)
39. Kiatpongsan, S., Tannirandorn, Y., Virutamasen, P.: Introduction to stem cell medicine. J. Med. Assoc. Thai. **89**, 111–117 (2006)
40. Mason, C., Dunnill, P.: A brief definition of regenerative medicine. Regen. Med. **3**, 1–5 (2008)
41. Daar, A.S.: The future of replacement and restorative therapies: from organ transplantation to regenerative medicine. Transplant. Proc. **45**, 3450–3452 (2013)
42. Christ, G.J., Saul, J.M., Furth, M.E., Andersson, K.E.: The pharmacology of regenerative medicine. Pharmacol. Rev. **65**, 1091–1133 (2013)
43. Agrawal, N., Dasaradhi, P.V., Mohmmed, A., Malhotra, P., Bhatnagar, R.K., Mukherjee, S. K.: RNA interference: biology, mechanism, and applications. Microbiol. Mol. Biol. Rev. **67**, 657–685 (2003)
44. Kurisaki, K., Kurisaki, A., Valcourt, U., Terentiev, A.A., Pardali, K., Ten Dijke, P., Heldin, C.H., Ericsson, J., Moustakas, A.: Nuclear factor YY1 inhibits transforming growth factor beta- and bone morphogenetic protein-induced cell differentiation. Mol. Cell. Biol. **23**, 4494–4510 (2003)
45. Jiang, Z.Y., Zhou, Q.L., Coleman, K.A., Chouinard, M., Boese, Q., Czech, M.P.: Insulin signaling through Akt/protein kinase B analyzed by small interfering RNA-mediated gene silencing. Proc. Natl. Acad. Sci. USA **100**, 7569–7574 (2003)
46. Hsieh, A.C., Bo, R., Manola, J., Vazquez, F., Bare, O., Khvorova, A., Scaringe, S., Sellers, W.R.: A library of siRNA duplexes targeting the phosphoinositide 3-kinase pathway: determinants of gene silencing for use in cell-based screens. Nucleic Acids Res. **32**, 893–901 (2004)
47. Zender, L., Hutker, S., Liedtke, C., Tillmann, H.L., Zender, S., Mundt, B., Waltemathe, M., Gosling, T., Flemming, P., Malek, N.P., Trautwein, C., Manns, M.P., Kuhnel, F., Kubicka, S.: Caspase 8 small interfering RNA prevents acute liver failure in mice. Proc. Natl. Acad. Sci. USA **100**, 7797–7802 (2003)
48. Zhou, T., Aumais, J.P., Liu, X., Yu-Lee, L.Y., Erikson, R.L.: A role for Plk1 phosphorylation of NudC in cytokinesis. Dev. Cell **5**, 127–138 (2003)
49. Li, Q., Brass, A.L., Ng, A., Hu, Z., Xavier, R.J., Liang, T.J., Elledge, S.J.: A genome-wide genetic screen for host factors required for hepatitis C virus propagation. Proc. Natl. Acad. Sci. USA **106**, 16410–16415 (2009)
50. Bric, A., Miething, C., Bialucha, C.U., Scuoppo, C., Zender, L., Krasnitz, A., Xuan, Z., Zuber, J., Wigler, M., Hicks, J., McCombie, R.W., Hemann, M.T., Hannon, G.J., Powers, S., Lowe, S.W.: Functional identification of tumor-suppressor genes through an in vivo RNA interference screen in a mouse lymphoma model. Cancer Cell **16**, 324–335 (2009)
51. Whitehurst, A.W., Bodemann, B.O., Cardenas, J., Ferguson, D., Girard, L., Peyton, M., Minna, J.D., Michnoff, C., Hao, W., Roth, M.G., Xie, X.J., White, M.A.: Synthetic lethal screen identification of chemosensitizer loci in cancer cells. Nat. **446**, 815–819 (2007)

52. Struwe, W.B., Warren, C.E.: High-throughput RNAi screening for N-glycosylation dependent loci in *Caenorhabditis elegans*. Methods Enzymol. **480**, 477–493 (2010)
53. Dominska, M., Dykxhoorn, D.M.: Breaking down the barriers: siRNA delivery and endosome escape. J. Cell Sci. **123**, 1183–1189 (2010)
54. Kong, J., Yu, X., Hu, W., Hu, Q., Shui, S., Li, L., Han, X., Xie, H., Zhang, X., Wang, T.: A biomimetic enzyme modified electrode for H_2O_2 highly sensitive detection. Analyst **140**, 7792–7798 (2015)
55. Ma, H., Yan, T., Zhang, Y., Gao, P., Pang, X., Du, B., Wei, Q.: A biomimetic mussel-inspired photoelectrochemical biosensing chip for the sensitive detection of CD146. Analyst **140**, 5019–5022 (2015)
56. Mascini, M., Macagnano, A., Monti, D., Del Carlo, M., Paolesse, R., Chen, B., Warner, P., D'Amico, A., Di Natale, C., Compagnone, D.: Piezoelectric sensors for dioxins: a biomimetic approach. Biosens. Bioelectron. **20**, 1203–1210 (2004)
57. Parry, M.A., Andralojc, P.J., Scales, J.C., Salvucci, M.E., Carmo-Silva, A.E., Alonso, H., Whitney, S.M.: Rubisco activity and regulation as targets for crop improvement. J. Exp. Bot. **64**, 717–730 (2013)
58. Jain, K., Mehra, N.K., Jain, N.K.: Potentials and emerging trends in nanopharmacology. Curr. Opin. Pharmacol. **15**, 97–106 (2014)

Biomimetics Strategies to Overcoming Noise

Syed W. H. Rizvi, Birgit Weyand, Meir Israelowitz, Christoph Gille, Matthias Reuter, Sabine Bohlmann, Kerstin Reimers, Peter Vogt, and Herbert P. von Schroeder

Abstract Noise is considered an artefact, in the practical use it is overcome by cool down the system in general at 77 K. We consider three biological examples which overcome noise by filtering the ratio between the signals and noise by using a distribution system. Noise is considered here in statistical terms poison, where the incident photon or otherwise in a detector and the detector has not influence and cannot increase.

Keywords Thermal noise · Poison · Signal/noise ratio optimization · Infrared · Terahertz and magnetic detectors

1 Introduction

Noise, as defined by Frieden [1], assumes that x_m is an event (for example a photon) where $P(y_m|x_n) = P(y_m)$, and P is the probability, y_m is the event passing through a filter or communication filters, by the law of largest numbers (i.e. the sample size gets closer to the average size of the whole sample).

Dereniak et al. [2] described noise using the analogy of a set of automobiles exiting a highway, and to be able to be detected, the automobiles need to exit at

S. W. H. Rizvi (✉) · M. Israelowitz · C. Gille · S. Bohlmann · H. P. von Schroeder
Biomimetics Technologies Inc., Toronto, Canada
e-mail: syed@biomimeticstechnologies.com

H. P. von Schroeder
University of Toronto, Toronto Western Hospital, Toronto, Canada

B. Weyand · K. Reimers · P. Vogt
Department of Plastic, Hand and Reconstructive Surgery, Hannover Medical School, Hannover, Germany

M. Reuter
Technical University Clausthal-Zellerfeld, Julius-Albert-Str. 4, Clausthal-Zellerfeld, Germany

M. Israelowitz et al. (eds.), *Biomimetics and Bionic Applications with Clinical Applications*, Series in BioEngineering,
https://doi.org/10.1007/978-3-319-53214-1_10

same speed and distance since the photons event are random, the only option of detecting them is by slowing them down by cooling systems (77 K).

While noise as found in near infrared is described by Holst [3], many factors affect the transmittance; factors like airborne particle affect the transmittance of any signal, this is much less in far infrared (Terahertz). For this reason, many applications like medical ones consider far infrared scanning systems. Other types of noise, found biological systems include magnetic field noise, which is produced by thermal motion of electrons (Johnson Current Noise) [4].

Nature gives us several examples of how noise overcome [5]. Biological systems evolve in the direction of minimizing cost and saving resources [6]. We are considering three natural examples that may solve the problem of noise in near infrared systems: Phyton, Middle Infrared: *Melanophila acuminata* Beetle and Magnetic Fields of *Magnetospillium magnetotactum*.

2 Methods

2.1 Snakes

Pytone were obtained with help Pittsburgh Herpetology Society and through privates' breeders. The snakes where kept in room temperature, under strict ethical conditions, and fed once every week [7–9].

2.2 Histology

Frozen sections of Pythonomie and Boniae sensor were cut 5.00 μm mounted on slides and stained with Azure methylene blue eosin dissolved in phosphate buffer solution which gives a nuclear purple colour [10–13].

2.3 Melanophila Acuminata

Melanophila acuminata were provided by Richard Westcott and Nathan Schiff. They had been collected from two locations: the Sandy River delta in Multnomah Co., Oregon and Olallie Lake in Jefferson Co., Oregon, in the Cascade Range at 1615 m elevation. The beetles were kept at 25 °C in a humidified environment and were fed raisins, peanuts and water.

2.4 Histology

Frozen sections of beetle sensory pit organs were cut at 5–10 μm thickness, mounted and stained with fuchsin Schiff, naphtol yellow and Sudan black [14] for differential staining of lipids [15–18], proteins [19] and polysaccharides.

Magnetospirillum gryphiswaldense was grown micro aerobically in flask standard medium [20] for 24 h at 25 °C as described earlier [21]. Cells were harvested by centrifugation (10,500 × g, 20 min, 4 °C) and washed twice with ice cold wash buffer (20 mM Hepes pH 7.4, 5 mM EDTA). Cell pellets were stored at −80 °C until use. Magnetosome isolation and purification with minor modifications was performed according to the protocol of Uebe et al. [22].

2.5 Scanning Electron Microscopy (SEM)

The SEM micrographs were taken with an electron microscope Phillips XL 30 FEG SEM. The electron microscopy samples were treated with ethanol to remove lipid in the cuticular region.

2.6 Zinc Phosphide

Single needles of zinc phosphide (Zn_3P_2) were grown by physical vapour transport [23] in a two zone furnace. Powder Zn_3P_2 (Sigma Aldrich) was used as the source material for needle growth. The material was [24] sealed under vacuum (<1 Pa) in quartz ampoules in which the growth took place, and were carbon coated by cracking of methane at 1000 °C in order to avoid chemical reaction between the Zn_3P_2 and the silica and prevent oxidation [25].

3 Results

Phyton has been described earlier in a morphological study by Weyand et al. [26]. The snake does not have a specific specialize sensor [9, 27–33], but Fig. 1 shows the system is just open ended nerves.

In the case of *M. acuminata* specialized sensor are present [34]. Figure 2 shows the sensor and Fig. 3 shows the sensor after special staining.

In the case of *magnetosomes* many examples can be considered; Fig. 4 shows the magnetosomes in *Magnetospirillum magnetotacticum* with flagellae used in orientation from a higher gradient to a lower oxygen concentration within magnetic fields [35].

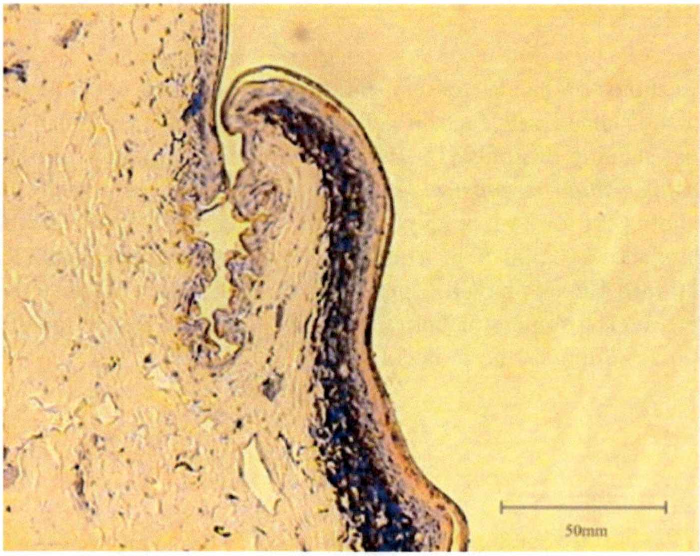

Fig. 1 Shows the open ended dendrites (where nerve activity occurs) which act as the sensor in python (via azure Methylene blue Eosin) [26]

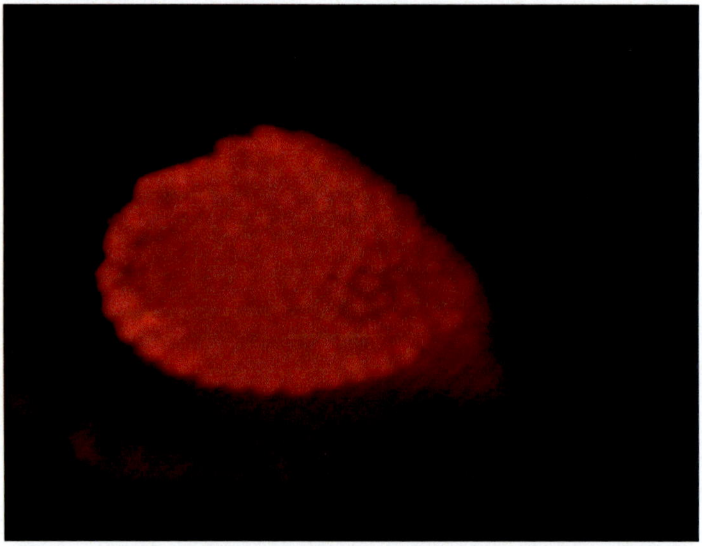

Fig. 2 Shows the sensor system of *M acuminata* under fluorescent light, consisting of 150–200 sensors

4 Discussion

Each of the biological examples have designed sensor systems that are adapted to reduction of noise, in some case more specialize than other just using what is available to achieved a functionality. In the case of Pythons, the IR detection

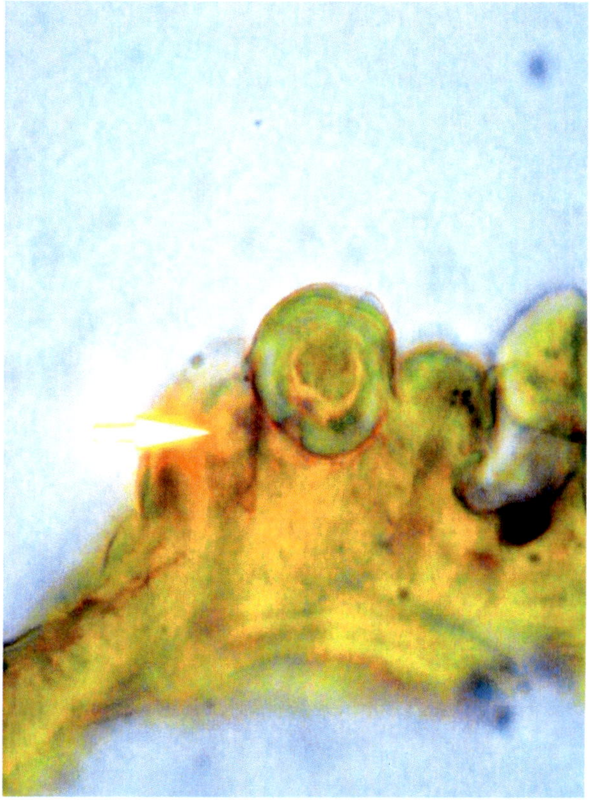

Fig. 3 Shows the sensor system of *M acuminata* after special staining. Staining denotes different components of the sensor, yellow is protein, and green is a mix of lipids with protein and red polysaccharide

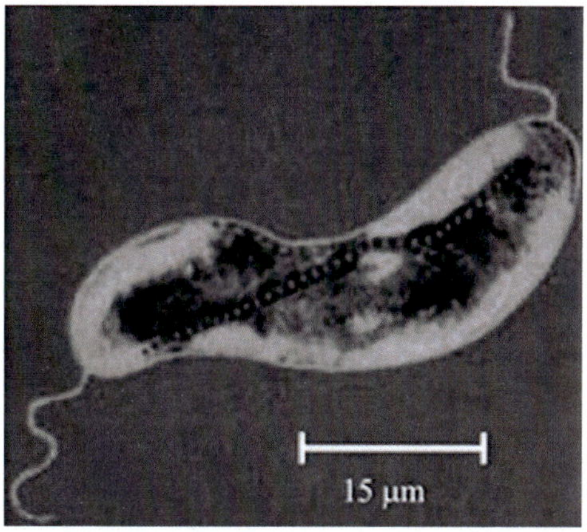

Fig. 4 Shows the magnetosomes in *Magnetospirillum gryphiswaldense* used to orient itself in magnetic field

mechanism is not made of photoreceptors—while photoreceptors detect light via photochemical reactions, the protein in the pits of snakes is a heat-sensitive ion channel (actually a temperature sensitive ion channel). It senses near-infrared signals through a mechanism involving warming of the pit organ, rather than chemical

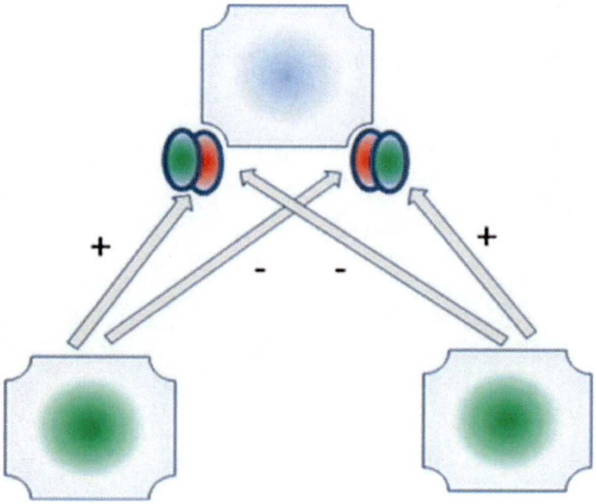

Fig. 5 Shows a binary inhibitor model with excitatory synapses

reaction to light [36]. *M. acuminata* has specialize sensors and the mechanism is photoreceptor-based [37], whereas in the case *magnetosomes* the system is based on magnetic field gradient [38].

The two options are possible to study Biomimetics strategies to overcoming noise: The first option utilizes the insertion of clamps for measuring signals originated from the activated sensor, and it is found in the biological process [7, 39]. The second option is to simulate the sensor using hybrid neural net structure, which involves two processing levels. The first level is for Signal/Noise ratio optimization done by so called DLS spectra and the second a biologically inspired visual signal processing unit (basic module) which models optical and acoustical pattern recognition in ear and eye; by a three-neuron-structure with INEX-synapses (Binary inhibitor model) as shown in Fig. 5. This basic module structure—which solves also the XOR-problem—is a high sensitive edge detector enabling the accentuation of even minimal contrasts in visual scenes [40–45].

This model works well with any level of signal input and when *M. acuminata* sensor is simulated it is shown to take an average of the signal, filter the noise since the sensor is continuously taken samples filtering the signal, which is possible to obtain information, a system to simulated based on specific process for example using Zinc phosphide [24, 46, 47] micro-wires [25, 48, 49], such system can be possible Fig. 6 show a single microwire and Fig. 7 shows the setup microchip setup (Fig. 8 shows the reader), with the signal reader and the same concept can be applied to *magentosomes* as show elsewhere [50, 51].

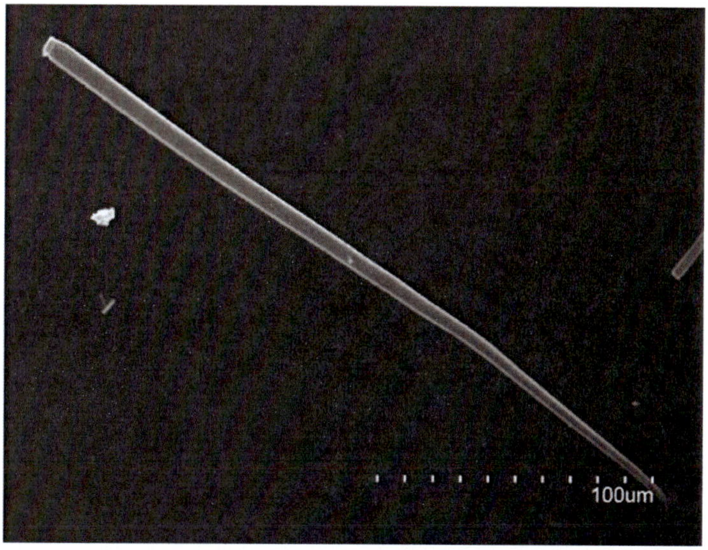

Fig. 6 Show a single Microwire as the building block to build the microchip, the diameter is 50 microns with a length of 100 micros

Fig. 7 Show the microchip and the space to set the microwires

Fig. 8 Inlet sensor readout and data processing

5 Conclusion

Distinct biological systems use the same principals to overcome noise by multiple distributed systems [52, 53], whether it is due thermal or photons, by having multiple sensors as a solution; this arrangement can average the ratio between the signal and noise. Noise as mention when the sample is sizes of the whole signal ratio crated the noise by continually picking the sample refined the signal in each case to achieve an objective, snake for pray, *M. acumainata* can detect fires from 150 km, using the burn wood, which is actually soft to deposited eggs, then larvae have protection and food. Bacteria case is anaerobic, moving from higher concentrations of oxygen versus lower concentration since the iron oxidises and sulphide iron oxide is found in this regions. The biological systems show a solution for far infrared, terahertz and magnetic detectors.

References

1. Frieden, B.R.: Probability, Statistical Optics and Data Testing. Springer, New York, NY (1991)
2. Dereniak, D.L., Boreman, G.D.: Infrared Detectors and Systems. John Wiley & Sons, New York (1996)
3. Holtst, G.S.: Common Sense Approach to Thermal Imaging. SPIE Optical Engineering Press, Winter Park, Florida US (2000)
4. Lee, S.K., Romalis, M.V.: Calculation of magnetic field noise from high permeability magnetic shields and conducting objects with a simple geometry. J. Appl. Phys. **103**, 084904 (2008)
5. Wolken, J.J.: Light Detectors, Photoreceptors and Imaging System in Nature. Oxford University Press, New York Oxford (1995)
6. Niven, J.E., Scharleman, J.P.W.: Do insects metabolic rates at rest and during flight scale with body mass? Biol. Lett. **1**, 346–349 (2005)
7. Waterman, T.H.: Animal Navigation. Scientific America Library, W. H. Freeman & Co., New York
8. Banks, R.C., McDiarmid, R.W., Gardner, A.L.: Checklist of Vertebrates of the United States, the U.S. Territories, and Canada, no. 166. Federal Government Series: Resource Publication (1987)
9. Bullok, T.H., Cowles, R.B.: Physiology of an infrared receptor: the facial pit of pit vipers. Science **115**, 541–543
10. Chailapakula, O., Wonsawat, W., Siangprohb, W., Grudpan, K., Zhaod, Y., Zhud, Z.: Analysis of sudan I, sudan II, sudan III, and sudan IV in food by HPLC with electrochemical detection: comparison of glassy carbon electrode with carbon nanotube-ionic liquid gel modified electrode. Food Chem. **109**, 876–882 (2008)
11. Lillie, R.D.: Conn's Biological Stains. Williams & Wilkins, Baltimore, MD., U.S.A
12. Baker, J.R.: Principles of Biological Microtechnique. Methuen, London (1970)
13. Lansink, A.G.W.: Thin layer chromatography and histochemistry of Sudan Black B. Histochemie **16**, 68–84 (1968)
14. Himes, M., Moriber, L.: A triple stain for deoxyribonucleic acid, polysaccharides and proteins. Stain Technol. **31**, 67–70 (1956)

15. Patterson, C.M., Kruger, B.J., Dalez, T.J.: Lipid and protein histochemitry of enamel of fluoride-effects of fluoride. Calcif. Tissue Int. **24**, 119–125 (1977)
16. Vogel, M.: Observations on the structure of Cystierci of Taenia solium and Taenia saginata (Cestoda: Taeniidae). J. Parasitol. **49**, 86–90 (1963)
17. Ornstein, L., Hudson, A.: Spectral matching of classical cytochemistry to automated cytology. J. Histochem. Cytochem. **22**, 453–469 (1974)
18. Lûsis, O.: The histology and histochemistry of development and resorption in the terminal oocytes of desert locust, Schistocerca gregaria. Quar J Micro Sci. **104**, 57–68 (1963)
19. Briand, L., Nespoulos, C., Huet, J.C., Takahashi, T., Pernollet, J.C.: Lingand binding and physico-chemical properties ASP2, a recombination odorant-biding protein from honeybee. E. J. Biochem. **268**, 752–760 (2001)
20. Uebe, R., Voigt, B., Schweder, T., Albrecht, D., Katzmann, E., Lang, L., Böttger, L., Matzanke, B., Schüler, D.: Deletion of a fur-Like gene affects iron Homeostasis and Magnetosome formation in Magnetospirillum Gryphiswaldense. J. Bacteriol. **192**, 4192–4204 (2010)
21. Schüller, D.: Formation of Magnetosomes in Magnectotactic bacteria. J. Mol. Microbiol. Biotechnol. **1**, 79–86 (1999)
22. Uebe, R., Henn, V., Schüler, D.: The MagA protein of Magnetospirilla is not involved in bacterial magnetite biomineralization. Bacteriol. **194**, 1018–1023 (2012)
23. Muñoz, V., Decroix, D., Chevy, A., Besson, J.M.: Optical properties of zinc phosphide. J. Appl. Phys. **60**, 3282–3288 (1986)
24. Decroix, D., Munoz, V., Chevy, A.: Growth and electrical properties of Zn3P2 single crystals and polycrystalline ingonts. J. Mater. Sci. **22**, 1265–1270 (1987)
25. Israelowitz, M., Weyand, B., Leiterer, C., Munoz, V., Martinez-Tomas, C., Herraiz-Llacer, M., Slowik, I., Beleites, C., Fritzsche, W., Krafft, C., Henkel, T., M Reuter, Rizvi, S., Gille, C., Reimers, K., Vogt, P., von Schroeder, H.P.: Biomimetic-inspired infrared sensors from Zn3P2 microwires: study of their photoconductivity and infrared spectrum properties. New J. Sci. (2014)
26. Weyand, B., Israelowitz, M., Reuter, M., Bohlmann, S., Rizvi, S.W., Gille, C., Vogt, P., von Schroeder, H.P.: Morphological study of the near-infrared pit sensor of the Python. In: Biomimetics, Bionic Applications, with Clinical Applications. Springer International Publishing Switzerland (2015)
27. Klein, M.C.G., Gorb, S.N.: Epidermis architecture and material properties of the skin of four snake species. R. Soc. Publ. (2012)
28. Krochmal, A.R., Bakken, G.S., LaDuc, T.J.: Heat in evolution's kitchen: evolutionary perspectives on the functions and origin of the facial pit of pitvipers (Viperidae: Crotalinae). J. Exp. Biol. **207**, 4231–4238 (2004)
29. Bullock, T.H.: Radiant heat reception in snakes. Commun. Behav. Biol. A **1**, 10–29 (1968)
30. Chiasson, R.B., Bentley, D.L., Lowe, C.H.: Scale morphology in Agkistrodon and closely related Crotaline Genera. Herpetologica. **45**, 430–438
31. Von Düring, M., Miller, M.R.: Sensory nerve endings of the skin and deeper structures. Academic Press, London (1979)
32. Bullock, T.H., Fox, W.: The anatomy of the infra-red sense organ in the facial pit of pit vipers. Q. J. Microsc. Sci. **98**, 219–223 (1957)
33. Newman, E.A., Gruberd, E.R., Hartline, P.H.: The infrared trigemino-tectal pathway in the rattlesnake and in the python. J. Comp. Neurol. **191**
34. Israelowitz, M., Rizvi, S.W., von Schroeder, H.P.: Fluorescence of the "fire-chaser" beetle Melanophila acuminata. J. Lumin. **126**, 149–154 (2007)
35. Pfeiffer, H.: Determination of anisotropy field distribution in particle assemblies taking into account thermal fluctuations. Phys. Sataus Solidi A. **118**, 295–306 (1990)
36. Gracheva, E.A., Nicholas, I.T., Ingolia, N., Kelly, Y.M., Cordero-Morales, J.M., Hollopeter, G., Chesler, A.T., Sánchez, E.E., Perez, J.C., J S Weissman, Davis, J.D.: Molecular basis of infrared detection by snakes. Nature **464**, 1006–1011 (2010)

37. Israelowitz, M., Kwon, K.A., Rizvi, S.W., Gille, C., von Schroeder, H.P.: Mechanism of infrared detection and transduction by Beetle Melanophila Acuminata in memory of Jerry Wolken. J. Bionic Eng. **8**, 129–139 (2011)
38. Strauss, S., Israelowitz, M., Weyand, B., Müller, R., Henkel, T., Shüler, D., Uebe, R., Rizvi, S., Gille, C., von Schroeder, H.P., Reimers, K.: Ferro oxyize magnetic-torsional angle from Magnetospirillum gryphiswaldense. In: Biomimetics, Bionic Applications, with Clinical applications. Springer International Publishing Switzerland (2015)
39. Evans, G.: Infrared receptors in Melanophila acuminata De Geer. Nature **202**, 211 (1964)
40. Reuter, M., Bohlmann, S.: Automatic detection of buried utilities in georeferenced multi-sensor data with neural networks. Presented at the TOK, Izmir, Turkey August (2011)
41. Reuter, M.: Computing with Activities V. experimental proof of the stability of closed self organizing Maps (gSOMs) and the potential formulation of neural nets. Presented at the WAC 2008, Waikoloa, Hawaii, USA (2008)
42. Reuter, M., Lenkl, K., Schroeder, O., Gramowski, A., Jügelt, K., Priwitzer, B.: Information extraction from biphasic concentration-response curves for data obtained from neuronal activity of networks cultivated on multielectrode-array-neurochips. Presented at the BMC Neuroscience January (2010)
43. Reuter, M.: Of the Stability of Closed Self Organising Maps (gSOMs) for Predictive Control. Presented at the, Lyon, France (2008)
44. Reuter, M.: Supervising cathodic protected gas nets with CI-based methods. Presented at the ISC'2013, 11th Annual Industrial Simulation Conference, Ghent, Belgium, 22 May 2013
45. Reuter, M., Bohlmann, S.: Supervising MultiCut Aggregates by Special Neural Nets. Presented at the WAC 2012, Puerto Vallarta, Mexico (2012)
46. Riedmiller, M., Braum, H.: A direct Adaptive Method for Faster Backpropagation Learning: The RPROP Algorithm, Neural Networks. Presented at the IEEE International Conference (1993)
47. Munoz, V., Decroix, D., Chevy, A., Besson, J.M.: Optical properties of zinc phosphide. Appl. Phys. **69**, 3282–3288 (1986)
48. Israelowitz, M., Rizvi, S.W., Holm, C., Gille, C., von Schroeder, H.P.: Method for producing a microchip that is able to detect infrared light with a semiconductor at room temperature
49. Israelowitz, M., Rizvi, S.W., Holm, C., Gille, C., von Schroeder, H.P.: Method to detect poor infrared rays, microchip that is able to detect poor infrared rays and apparatus working with these microchips
50. Israelowitz, M., Rizvi, S.W., Gille, C., Holm, C., von Schroeder, H.P.: Method for detection of poor sources of electrical and magnetic fields
51. Israelowitz, M., Rizvi, S.W., Gille, C., Holm, C., von Schroeder, H.P.: Microchip for the detection of poor sources of electrical and magnetic fields
52. Titterington, D.M.: Bayesian methods for neural networks and related models. Stat. Sci. **19**, 128–129 (2004)
53. Kaski, S.: Data exploration using self-organizing maps. Acta Polythecnica Scand. Math. Comput. Manag. Publ. Finn. Acad. Technol. 57–60 (1997)

Computer Models in Biomimetics

Biological Inspired Optical Pattern Analysis by Topological Neurons

Matthias Reuter and Sabine Bohlmann

Abstract Common optical pattern analysis can be characterized by diverse sophisticated mathematical methods, similar to the known mechanism of neural activities. It is known that living primitive creatures with small numbers of neurons can recognize objects, navigate in their environment and distinguish between similar objects. Based on our investigations regarding the neural coding, storing and retrieving information and topological structures of the dendrite trees, we developed a new theory about the pure neural-based optical signal analysis. Basic assumptions of this theory are that neurons only fire or not fire, that they are combined with each other and that their synaptic structures seems to follow topological structures. These assumptions have not been taken into account by other models. Together which the cybernetic model developed by Helmuth Benesch's "Carrier-Pattern-Meaning" [1], we developed and tested a simple algorithm to find and highlight structures, textures, and hidden or disturbed objects from random pictures. We have found that the overall basic principle of our algorithm can be defined as a potential oriented approach to describe the neuronal activity and interaction.

Keywords Neural networks · Simulation of Eyes information pathways · Synaptic pathways · Radargram analysis

1 Introduction

Students of neural nets know about the so-called XOR-problem regarding the linear separability of a feature space by simple perceptrons and that it can be handled only by Multilayer Perceptron architectures (MLPs) [2, 3]. We have learned from biology that linear separability is a fundamental part of the early pattern recognition procedures, and that complex architectures like MLP are neither found in the

M. Reuter · S. Bohlmann (✉)
Department of Big Data and Technical Information Systems, TU Clausthal - Institute for Informatics/IngB RT&S, Clausthal-Zellerfeld, Germany
e-mail: sabine.bohlmann@tu-clausthal.de

M. Israelowitz et al. (eds.), *Biomimetics and Bionic Applications with Clinical Applications*, Series in BioEngineering,
https://doi.org/10.1007/978-3-319-53214-1_11

cochlea nor in the retina. It was this unsolved problem that led us to develop a new theory regarding neuron and dendrite structure to enable XOR-like signal processing structures to detect edges and enlarge the contrast. As these structures are not conditionable, and given that one and only one neural layer occurs in those neurological structures, it was clear for us that nature has solved the XOR-problem by this special architecture.

The basis of this special architecture is the terminal arborisation of an axon [4]. It is important to mention that this arborisation stands in contrast to the common simulated neural network theory; axons of biological neurons split in several branches. In that way it is possible for one axon to simultaneously innervate dendrites of different neurons. The synapses of the axon terminals can act either in an inhibitory or excitatory way which means that one and only one axon can stimulate or suppress the activity of innervated neurons [5].

2 Models

2.1 Theory and Used Models

For a neural net structure solving the XOR-problem we use a very simple structure, shown in Fig. 1. On layer one we have the input layer, on layer two the output layer. The axon of each input layer neuron in our model has a two way excitatory terminal arborisation: one ends at the dendrite synapse of the classification neuron, the other ends on the dendrite synapse of an inter-neuron [6, 7]. This inter-neuron is converting the information of the input neuron and forwarding this converted information via its axon to the opposite synapse of the classification neuron. In this way an excitatory and an inhibitory contribution is acting at every synapse of the classification neuron. This is symbolized by two ellipsoids at the synaptic structure of the neuron. Furthermore the inhibitory pathways is characterized by "–", the excitatory pathway by "+".

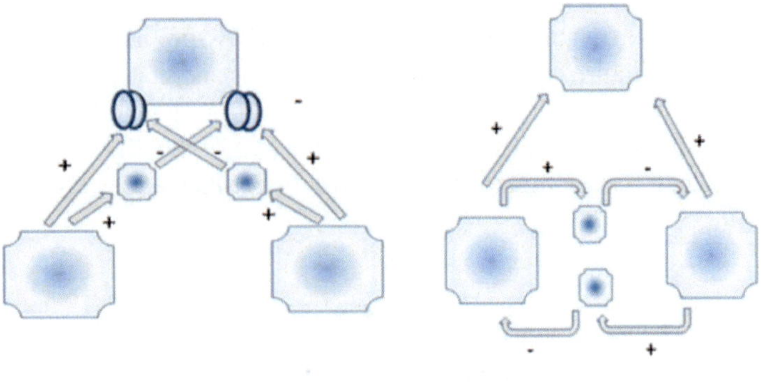

Fig. 1 a, b Structures of XOR-problem solving neural architectures including inter-neurons

For a better understanding, we now change the model by dropping the inter-neurons and visualising only two axons from input neuron acting at the synapses of the classification neuron. In Fig. 2a the arborisation is realized in such a way that on one side the output neuron is innervated by both input neurons (as in every common simulation), but on the other side each input neuron innervates the other input neuron via an inhibitory connection. Such so-called direct lateral inhibitory structures can be found both in the retina (eye) and cochlea (ear). In Fig. 2b the arborisation is realized in such a way that every input neuron innervates the synapse of the other input neuron in an inhibitory way. This inhibition structure is called direct backward inhibition.

Next, using Table 1, we will calculate the different result of our two networks.

Now we assume that the resulting activities of the input neurons are the sum of all inputs; both input layer neurons have the same activity values and the neurons have a Heaviside step function [8, 9] as an activation, shown in Fig. 3.

Now we are enlarging our model by the principle “Carrier-Pattern-Meaning”, meaning that the neurons are the material carrier. Their state of activation, the activity pattern and their “active/not active” status all have construed meaning. In detail we visualized this principle in the following way: if a neuron is active it is a green light, whereas inhibitory activity is a red light.

To evaluate the model we start with the inhibitory structure of the input layer. In Fig. 4 both cases a and b: no input in both input neurons and equal input in both input neurons are shown.

It can be easily pointed out that the desired output of the output neuron is realized by the neuron structure shown in Fig. 2b, meaning that in both cases the resulting activity of the output neuron is equal to zero.

In Fig. 5 the other two possible activity patterns of the input layer are visualized. It is the Heaviside step function which guarantees the desired activity “1” of the output neuron.

Returning to the left hand model of Fig. 2, where we used a two way synaptic structure: we start again with no- and overall activity of the input layer neuron (Fig. 6). Furthermore, we assume that the resulting activity of the both output layer synapses is the sum of the inputs of the terminal arborisation contributions.

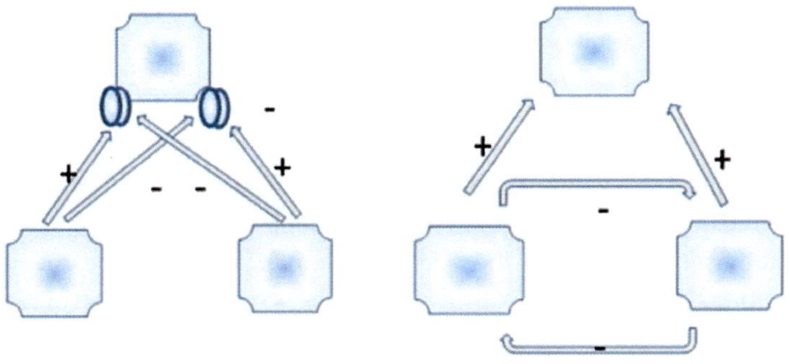

Fig. 2 **a**, **b** Possible structures of simulated neural networks, solving the XOR-problem

Table 1 XOR-like structure

Activity of the output neuron	Activity of the left input neuron	Activity of the right input neuron
0	1	1
1	0	1
1	1	0
0	0	0

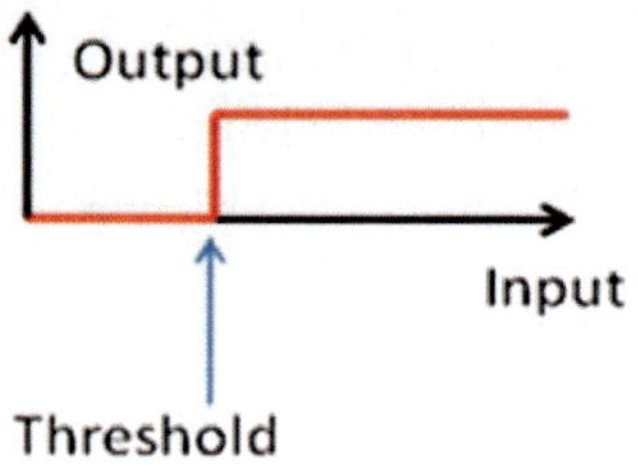

Fig. 3 The Heaviside step activation function (HSF)

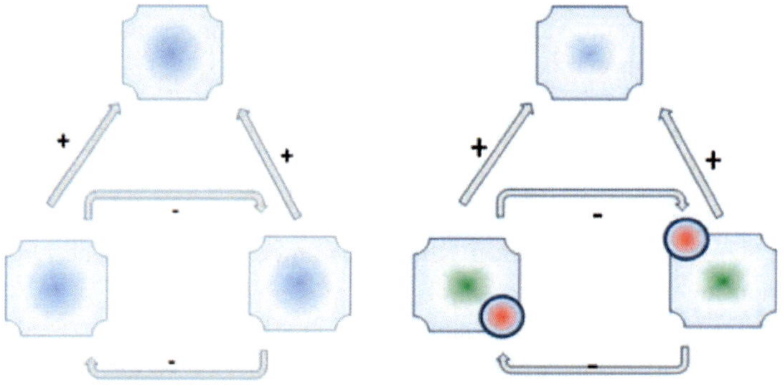

Fig. 4 Activity of the output neuron with inhibitory input layer structure. **a** No input in the input layer, **b** overall input in the input layer

It is important to again use the Heaviside step function for each synapse, to ensure that the different excitatory and inhibitory contributions of the axons are combined in a linear way. In Fig. 7a and b the other two possible activity pattern of the input layer are visualized.

The result of our evaluation shows that both structures solve the XOR-problem by a 3-neuron-architecture. We call such architectures including the three-neuron-model handling the XOR-problem **XOR-structures**. Neurons where an excitatory and inhibitory pathway fit at the same synaptic structure are **topological oriented neurons**, or **TO-Neurons**. The pre- and postsynaptic structure where the excitatory and inhibitory axon-terminals are involved is called the **INEX-Synapse** (Fig. 8).

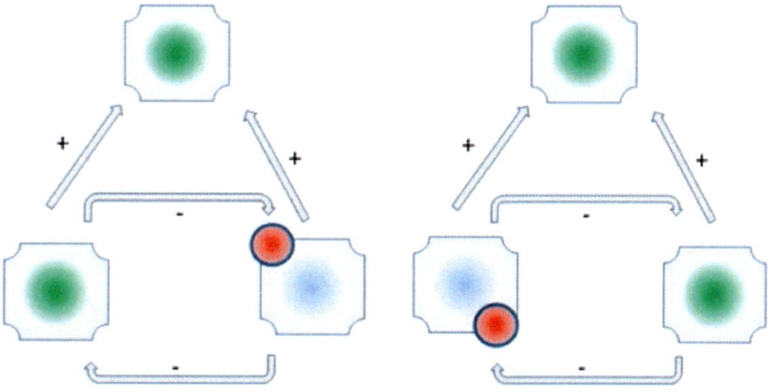

Fig. 5 Activity of the output neuron if an inhibitory input layer structure is given. **a** Activity of the right hand side neuron is larger than null. **b** Activity of the left hand side neuron is larger than null

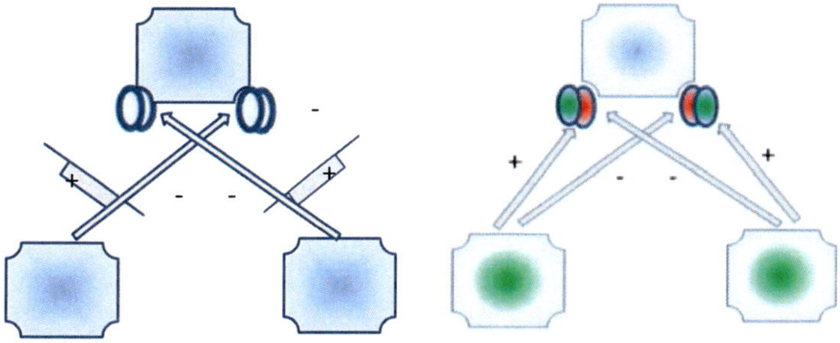

Fig. 6 Activity of the output neuron if a complex synaptic structure is given. **a** No input in the input layer is given. **b** Overall input in the input layer is given

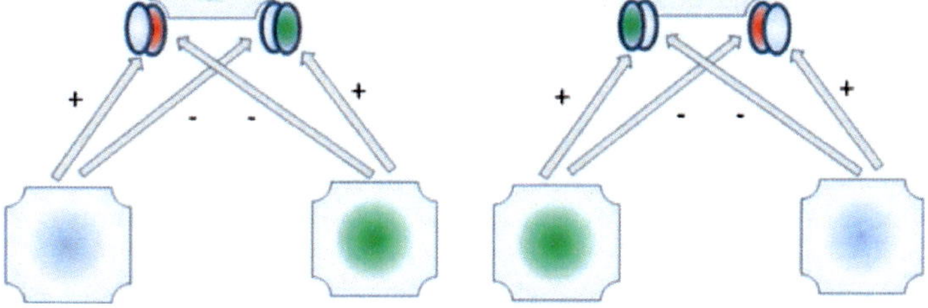

Fig. 7 Resulting activity of the output neuron if a two way synapse structure is given. **a** Right hand side activity of an input neuron. **b** Left hand side activity of an input neuron [10]

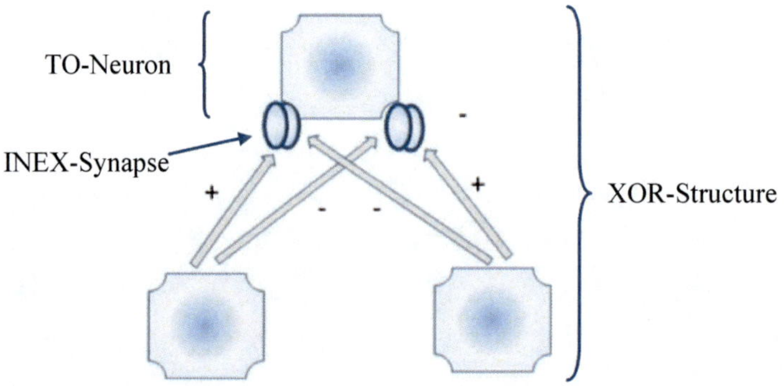

Fig. 8 Formalized XOR-structure

Fig. 9 In a defined interval linear activation function (IAF)

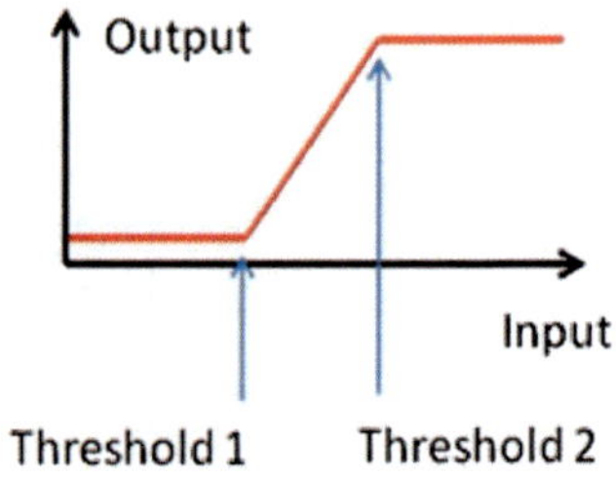

One advantage of TO-Neurons is to use the topological characteristic [11, 12] of the synapses to detect gradients like slightly intensity changes or edges when we replace the Heaviside step function of the synapses by a (in a defined interval) linear activation function [13, 14] (see Fig. 9).

To detect the direction of the gradient and therefore determine if the intensity is increasing or decreasing respectively, and to define the edge-form, we have to define a time structure of the input neuron's activity by a chronology of its synaptic activities.

A form of such a chronology can be realized as follows: first we define the coding of the resulting neuronal activity transmitted via an axon: The energy evokes at the axon hillock a time series of action potentials whereby the number of potentials are a function of the power of the activity pattern [15, 16]. If we do so, the neurons activity structure will have a time structure, which is similar to spike-coding known from biological neurons [17–19], whereby we assume that large input values are coded by a high rate of spikes and low input vales by a low rate of spikes (see Fig. 10).

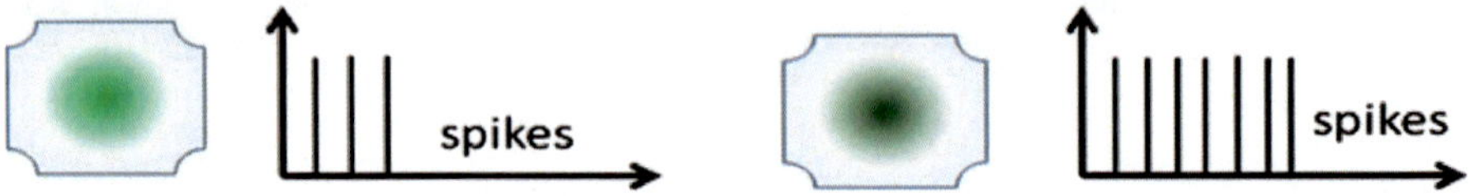

Fig. 10 Spiking behaviour of TO-neurons: left low power of neuron's activity, right high power of neuron's activity

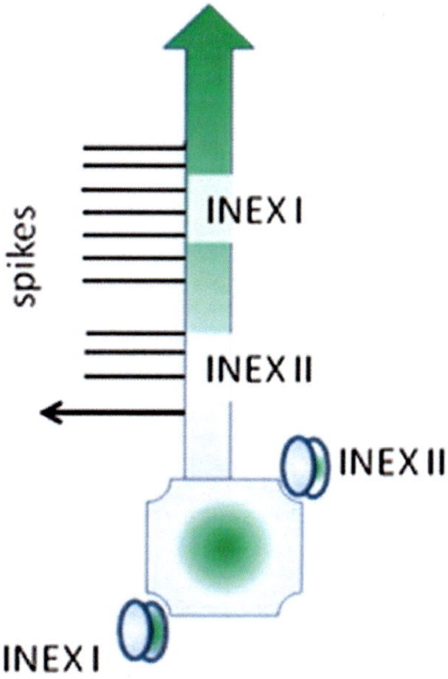

Fig. 11 Chronological response evoked by an asymmetric INEX-structure

Secondly we define a topological order of the synapses of the input neuron shown in Fig. 11.

Let's say the INEX-Synapse I is placed at the left side at the bottom of the TO-Neuron while the INEX-Synapse II is placed at the right side at the top of the TO-Neuron. Assuming that the time which is necessary for the potential propagating to the axon hillock is a function of the distance: 'INEX-Synapse—Axon hillock' we can expect that INEX-Synapse II evokes spiking behaviour earlier than INEX-Synapse I. In that way right synapse activity is represented by the first part of a neurons spiking activity pattern and the activity of the left synapse is represented by the second part of the spiking activity pattern.

Defining that the resting potential of our TO-Neuron is equal "−1" and the action potential is "1" Fig. 12 points out the potential structures of light left side and light right side.

Finally, we have to find an adequate interpretation of the TO-neurons activity with their (in a defined interval) linear activation function and the activity modus of the adjacent cells (e.g. ganglion cells). We start with physics, chemistry and logic: as the retinal cells acts in a pure electrical way, and the synapses of the adjacent cells in a pure chemical way, and the act of recognition in the lower parts of the visual system are following both ways, the information processing has to be in the same way [20]. The only modality we know is the interpretation of both as potentials. Potentials act either as summative or subtractive. With this, we have found the mathematical operations for use when building complex neural structures to analyse optical patterns regarding structures, textures or noisy contributions.

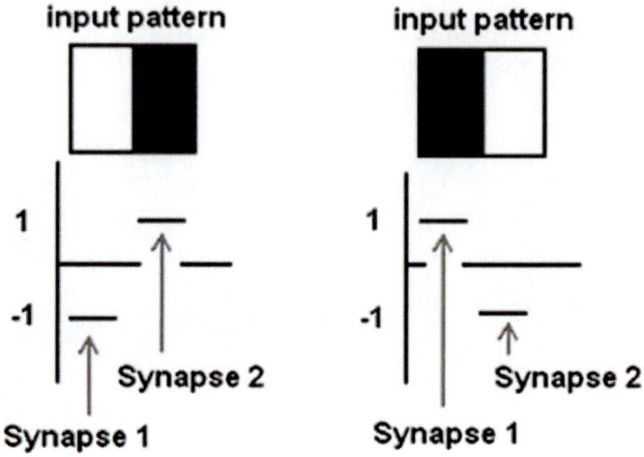

Fig. 12 Activity for synapse 1 and synapse 2 for right hand side (**a**) and left hand side edge (**b**)

3 Results

3.1 Application

In this section we first discuss a special application of the TO-Neurons to show that our biologically-inspired structures are very helpful mechanisms for rapid pattern recognition algorithms.

If we apply TO-Neurons, we first consider the parameters underlying different kinds of pattern analysis routines. Out of our model we found that there are only four parameters for use:

- Change the activation function
- Change the thresholds of the activation functions
- Change the network the TO-Neurons are involved in
- Change the threshold for the spike generation (in the axon hillock).

Remark: The last parameter will not be taken into account here.

3.2 Activation Functions and Thresholds Used for Pattern Recognition

In Figs. 3 and 9 we introduced two kinds of linear activation functions. Both are simple, but both lack the fact that on the left and right side of their linearity the neurons are more or less "blind" for every kind of input change. This phenomenon is called "Grossbergs saturation theorem" [21]. To guarantee that the neurons activity will follow all kinds of input changes—surely in a given activation interval—we can use a kind of shifted hyperbolic tangent, exemplary shown in Fig. 13.

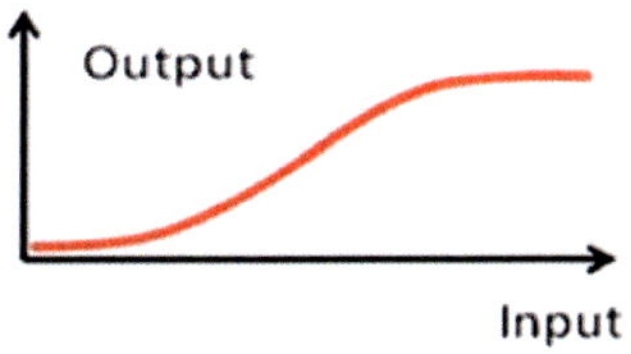

Fig. 13 Smooth activation function (SAF)

As the SAF-function is smooth over the whole positive input and output interval no threshold parameters can be found. Otherwise we can change the slope of the function via a new kind of parameter which results at least in the increasing or decreasing of the sensitivity of the TO-Neurons.

For better understanding, Fig. 14 shows two XOR-structures with different slope factors: on the left the slope factor is 1, and on the right the slope factor is 0.02. In the second case the lower slope results in a lower activation of the top TO-Neuron; which means, as shown here that the activated neuron of the lower structure has to be more activated to evoke the same activation in the upper neuron.

- This behaviour can be used to sensitize the pattern analysis, as it holds for the slope factor.
- The larger the slope factor the smaller the operating range of the pattern recognition
- The smaller the slope the larger the operating range of the pattern recognition.

In our software application tools, the operator can select one of these activation functions and the slope parameter. Through this we found that it depends on the form and structure of the pictures to be analysed which activation function and/or slope factor carries out an optimal recognition and/or signal analysis.

Fig. 14 XOR-structures with different activation slope of SAF. Left hand side is the slope factor is equal 1, right hand side the slope factor is 0.02

4 Discussion

4.1 Operators Combining the XOR-Structures

4.1.1 Basic Operators Defining Receptive Fields

The use of the XOR-Structures implies that we have to define how many neurons of the structure are combined and the way in which they are combined to each other. This combination is more or less nothing else than defining which kind of receptive field dimension should be used in a pattern recognition process [22, 23]. As we know from biology in the fovea centralis we find very small receptive fields, in its periphery the receptive fields become larger. It is important to mention, that the smallest receptive field as least needs two sensitive neurons (photoreceptors) to detect a gradient means a light evoked activity change. So the simplest structure of a receptive field is the XOR-Structure itself.

If a simple XOR-Structure is used, only on direction of gradient detection can be done. The direction itself is defined by the topological placement of the XOR-Structure in the retina.

If a 360° edge detection should be done by one XOR-Structure the receptive field has to be symmetrically as shown exemplary in Fig. 15. For a better clarity only three of the nine inputs are visualized.

For our investigation we used several receptive field structures which we call *Field-Operators*. In Fig. 16 the most important of them are shown.

In our software application tool, the operator can select one of these Operators, whereby we found that it depends on the form and structure of the pictures to be analysed which Field- Operator carries out an optimal recognition and/or signal analysis.

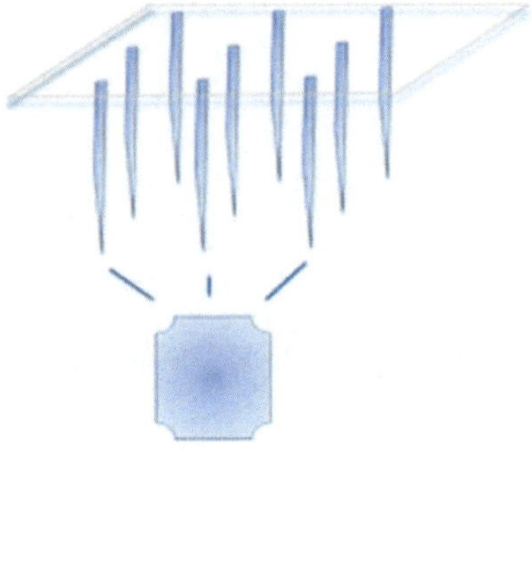

Fig. 15 360° view XOR-structure, called operator 1 (Op1)

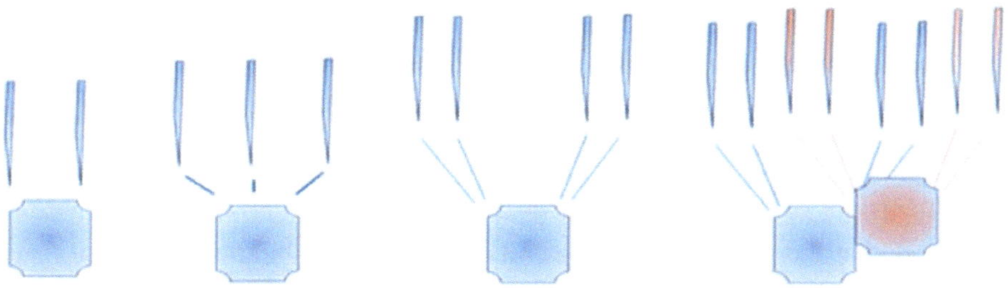

Fig. 16 Forms of the basic Field-Operators—resulting from the fundamental XOR-Structure. From left to right: Operator 2 (Op2), Operator3 (Op3), Operator 4 (Op4), Operator 5 (Op5)

4.2 Operators Defining Lateral Inhibition Routines

Beneath the selection of the field operators, for pattern recognition the normalisation regarding the amplitude of the different picture pixel are of fundamental importance. If our approach that the XOR-Structures represent the basic module of biological oriented optical pattern recognition, the principal functionality of a XOR-Structure should handle the normalisation as well. Looking again to the things we know about the structure of the retina we easily can identify the lateral inhibition done by horizontal cells as the operation which performs the normalisation regarding the amplitude [24, 25]. In principle this lateral inhibition can be schematically pointed out as shown in Fig. 18. The different contributions of the photoreceptors are collected in the horizontal cell which calculates the amount of inhibition. This amount is reverted back to the photoreceptors which are inhibited. In detail the amount of inhibition is nothing else than the gradient of the smooth activation function (see Fig. 17) whereby the left shift correspondence to the amount of inhibition of the horizontal cell and the downshift to resulting inhibition of the photoreceptor.

This consideration applied, the following structure of the lateral inhibition seems to be intuitive.

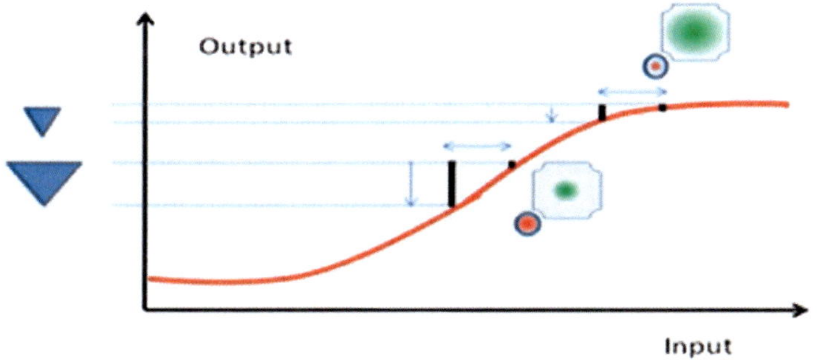

Fig. 17 Amounts of inhibition depending on strength of illumination

The model of the lateral inhibition shown in Fig. 18 should be interpreted as follows: the horizontal cell (blue cell at the bottom of Fig. 18) is connected to four photoreceptors. Out of the activity contributions of every photoreceptor the horizontal cell calculates its contribution of inhibition, whereby this contribution doesn't follow a linear trend but the SAF. The amounts of inhibition are symbolized in Fig. 18 by the different sizes of the triangles.

Note that the principle of the lateral inhibition is nothing but a type of floating average procedure similar to the so called Difference Auto Power Spectra (DLS-Spectra) described by Reuter [26]. These spectra have been used for many years to clean up random data from noise.

By defining the LI-operators we have found all necessary algorithmic structures to perform the XOR-Structure based optical pattern analysis. Giving a resume to the theoretical part, Fig. 16 points out on the right side that operations have to be adapted via the discussed LI-Operators in that way that the receptor and XOR-Structures act in an optimal way.

4.3 Example Applications

As we have shown the basic principle of the XOR-Structures we derived from neurology, physics and chemistry seems to be logical, but at least experiments have to show, if the algorithm really confirm the theory. Therefore, in this chapter we demonstrate how powerful our model is.

Figure 19 shows the Graphical User Interface (GUI) of our software tool. The software enables us to apply different neuron structures, changing intensity parameters and visualisation parameters, so at least the LI-Operators and the intensity of the internal visualization routines of the software tool.

Beneath those functionalities common loading, saving, databank routines to store and recall the neuronal parameters etc. are involved in the GUI.

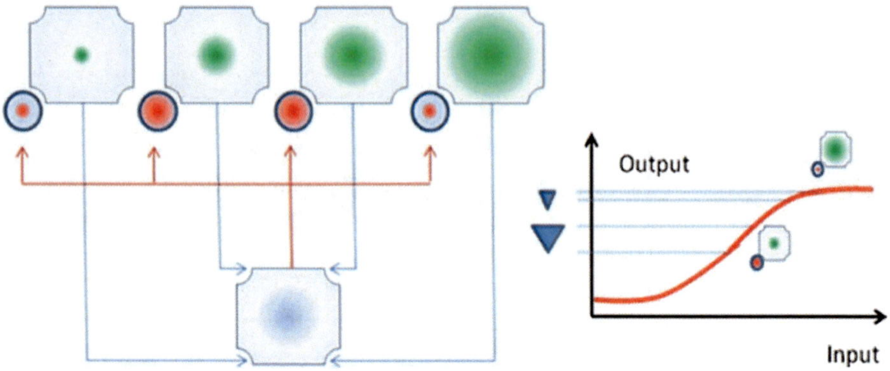

Fig. 18 XOR-structure based operator handling a lateral inhibition (LI-Operator 1)

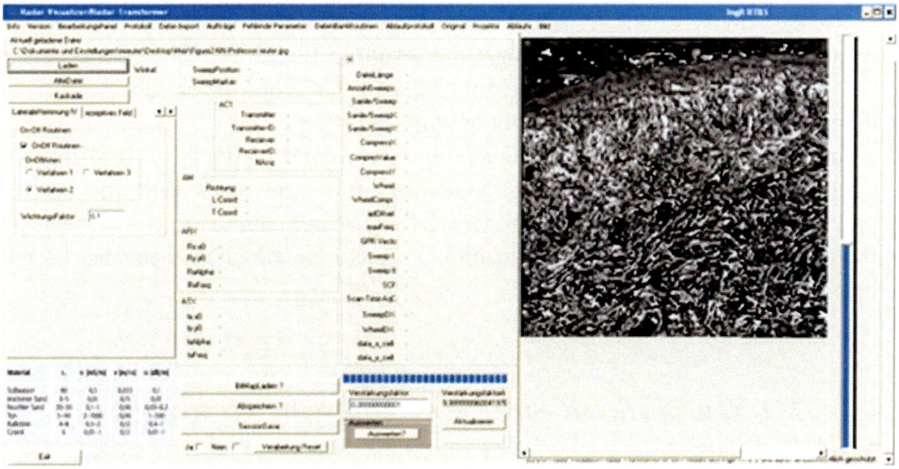

Fig. 19 Graphical User Interface of a TO-Neurons based picture analyse tool

4.4 Edge Detection by Common and TO-Neurons

In nearly every optical recognition routine one of the first processing steps is to detect the edges of the different objects of a picture. This normally is done by Sobel operators [27] for the x- and y-directions, given as follows:

$$
\begin{aligned}
X[3][3] = \\
& \{ \\
& \quad \{-1 \quad 0 \quad 1\}, \\
& \quad \{-2 \quad 0 \quad 2\}, \\
& \quad \{-1 \quad 0 \quad 1\} \\
& \}
\end{aligned}
$$

$$
\begin{aligned}
Y[3][3] = \\
& \{ \\
& \quad \{-1 \quad -2 \quad -1\}, \\
& \quad \{0 \quad 0 \quad 0\}, \\
& \quad \{1 \quad 2 \quad 1\} \\
& \}
\end{aligned}
$$

These operators acts point by point as convolution. In Fig. 20a the result of such a convolution on a ground penetrating radargram is shown. The white squares point out the hyperbolas of a buried object in the underground and the black rake the centre of gravity of the hyperbola. If we apply the TO-Neurons/XOR-Structures structure to the same picture, we get the resulting Fig. 20b.

As result we state that both operations lead to more or less the same results, only the centre of gravity of the hyperbola slightly differ.

In detail the XOR-Structure Operator 1, combined with the lateral inhibition operator 2 forms the underlying XOR-Structure. The black rake the centre of gravity—and therefore the real hyperbola detection/identification have been done by similar biological inspired algorithmic structures, which will not be discussed here.

4.5 TO-Neurons/XOR-Structures Analysing Histological Pictures

In Fig. 21 we show the pattern analysis of two histological pictures from part of a snake nervous system. The original tissue histology is placed on the left side.

5 Conclusion

In this chapter we show that a small change in simulated neurons simulation leads to totally new features in the neurons behaviour. Not only can the XOR-problem be solved by the combination of excitatory and inhibitory parts of two way terminal arborisations, but also a simple edge, and more generally gradient detectors result from this model enhancement.

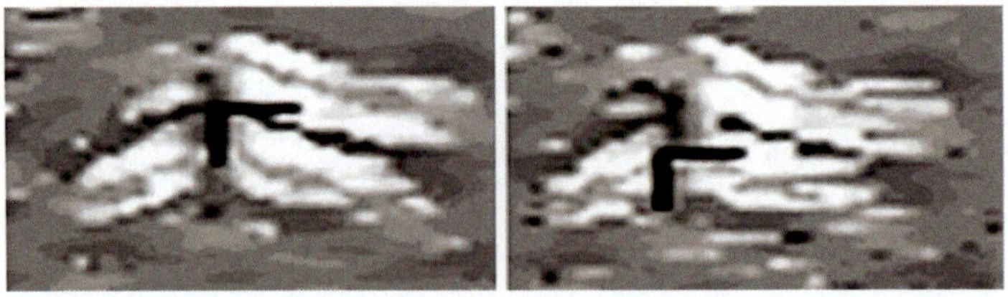

Fig. 20 Edge detection by **a** Sobel operators (left) and **b** TO-Neurons (right)

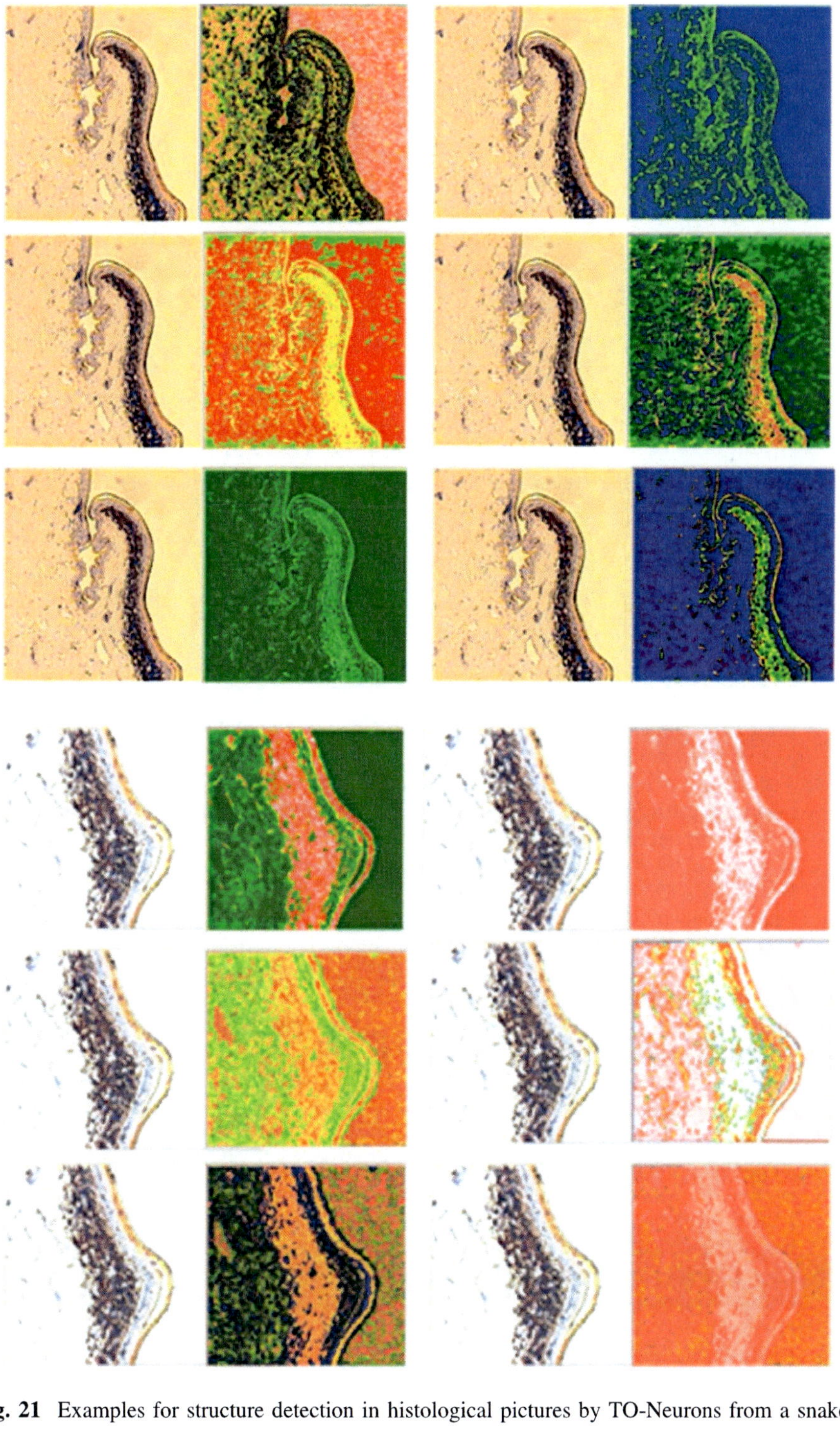

Fig. 21 Examples for structure detection in histological pictures by TO-Neurons from a snake

References

1. Benesch, H.: Der Ursprung des Geistes. dtv 1542 (1980)
2. Minsky, M.L., Papert, S.: Perceptrons. an introduction to computational geometry, 1st edn. MIT Press, Cambridge, Massachusetts (1968)
3. Rumelhart, D.E., Hinton, G.E., Williams, R.J.: Learning internal representations by error propagation. In: Rumelhart, D.E., McClelland, J.L. (eds.) Parallel Distributed Processing: Explorations in the Microstructure of Cognition, vol. 1. MIT Press, Cambridge, MA (1986)
4. Terence, R.A.: Neuroanatomy and the Neurologic Exam: A Thesaurus of Synonyms, Similar-Sounding Non-Synonyms, and Terms of Variable Meaning, p. 371. CRC Press. ISBN 978-0-8493-8631-2 (1994)
5. Fino, E., Packer, A.M.: The logic of inhibitory connectivity in the neocortex. Neuroscientist **19**(3), 228–237 (2013) (first published on 24 Aug 2012)
6. Labrakakis, C., Lorenzo, L.-E., Bories, C., Ribeiro-da-Silva, A., De Koninck, Y.: Inhibitory coupling between inhibitory interneurons in the spinal cord dorsal horn. Mol. Pain **5**, 24 (2009)
7. Kelsom, C., Lu, W.: Development and specification of GABAergic cortical interneurons. Cell Biosci. **3**, 19 (2013)
8. Burrows, B.L., Colwell, D.J.: The fourier transform of the unit step function. Int. J. Math. Educ. Sci. Technol. **21**(4), 629–635 (1990)
9. Heaviside, O.: Electromagnetic theory, vol. 1–3 (Classic Reprint). Forgotten Books (2010)
10. Bohlmann, S., Reuter, M.: Analysing radargrams and sonargrams by biological inspired signal processing methods to optimize the detection of mines and dumped ammunition. 10th Future Security Research Conference, Berlin, September 15–17 (2015)
11. van Elburg, R.A.J., van Ooyen, A.: Impact of dendritic size and dendritic topology on burst firing in pyramidal cells. PLoS Comput. Biol. **6**(5), e1000781 (2010). https://doi.org/10.1371/journal.pcbi.1000781
12. Van Ooyen, A., Van Elburg, R.A.J.: Dendritic size and topology influence burst firing in pyramidal cells. In: Cuntz, H., Remme, M.W.H., Torben-Nielsen, B. (eds.) The Computing Dendrite. Springer, New York, pp. 381–395 (2014)
13. Hopfield, J.J., Tank, D.: Neural computation of decisions in optimization space. Biol. Cybern. Nr. **52**, 141–152 (1985)
14. Churlchland, P., Sejnowski, T.: Grundlagen zur Neuroinformatik und Neurobiologie. Friedr. Vieweg & Sohn Verlagsgesellschaft, Braunschweig/Wiesbaden (1997)
15. Jack, J.J.B., Noble, D., Tsien, R.W.: Electric Current Flow in Excitable Cells. Clarendon Press, Oxford (1975)
16. Stuart, G., Schiller, J., Sakmann, B.: Action potential initiation and propagation in rat neocortical pyramidal neurons. J. Physiol. **505**(3), 617–632 (1997)
17. Masuda, N., Aihara, K.: Bridging rate coding and temporal spike coding by effect of noise. Phys. Rev. Lett. **88**(24), 248101 (2002)
18. Brown, E.N., Kass, R.E., Mitra, P.P.: Multiple neural spike train data analysis: state-of-the-art and future challenges. Nat. Neurosci. **7**, 456–461 (2004)
19. Restivo, L., Vetere, G., Bontempi, B., Ammassari-Teule, M.: The formation of recent and remote memory is associated with time-dependent formation of dendritic spines in the hippocampus and anterior cingulate cortex. J. Neurosci. **29**(25), 8206–8214 (2009)
20. Lam, D.M.K.: Synaptic chemistry of identified cells in the vertebrate retina. In: Cold Spring Harbor symposia on quantitative biology, vol. 40. Cold Spring Harbor Laboratory Press (1976)
21. Grossberg, Rami: Classification theory for abstract elementary classes. Contemp. Math. **302**, 165–204 (2002)
22. DeAngelis, G.C., Ohzawa, I., Freeman, R.D.: Receptive-field dynamics in the central visual pathways. Trends Neurosci. **18**(10), 451–458 (1995)

23. Henry, Geoffrey H.: Receptive field classes of cells in the striate cortex of the cat. Brain Res. **133**(1), 1–28 (1977)
24. Blakemore, C., Carpenter, R.H., Georgeson, M.A.: Lateral inhibition between orientation detectors in the human visual system. Nature (1970)
25. Cook, P.B., McReynolds, J.S.: Lateral inhibition in the inner retina is important for spatial tuning of ganglion cells. Nat. Neurosci. **1**(8), 714–719 (1998)
26. Reuter, M.: Analysing the structure of poly-crystalline materials by 2-dimensional DLS-spectra and neural nets. In: Computational Intelligence, LNCS 2206, pp. 420–427. Springer (2001)
27. Scharr, H.: Optimale Operatoren in der Digitalen Bildverarbeitung. Dissertation, Ruprecht-Karls- Universität, Heidelberg (2000)

Neural Networks for Modeling Metabolic Pathways

Meir Israelowitz, Birgit Weyand, Sabine Bohlmann, James Kramer, Christoph Gille, Syed W. H. Rizvi, Herbert P. von Schroeder, and Matthias Reuter

Abstract Metabolic pathways are enzyme-catalyzed reactions that can be described by the Michaelis-Menten equation. The inherent problem with Michaelis-Menten kinetics is that the model is not reversible. Reverse catalysis, which is common in metabolic pathways, can be better described by Cell Communication Protocol© which offers a Self-Organizing Map (SOM) for the visualization of the enzyme-catalyze reactions. Neural gas, an algorithm for finding optimal data, is added to the SOM and allows for an increase in the weight space to handle the potential complexity of enzyme-catalysis with the use of back propagation, the system allows for the modeling of reverse reactions.

Keywords Michaelis-Menten kinetics · Cell communication protocol · Self-organizing map · Neural gas · Back propagation · Metabolic pathways

1 Introduction

Analytic models have limitations when applied to biological applications, because the of both the simplicity and the complexity of the biological systems [1, 2]. *Neural networks* have provided a functional alternatives, since they can allow for

M. Israelowitz · C. Gille · S. W. H. Rizvi · H. P. von Schroeder
Biomimetics Technologies Inc, Toronto, Canada

H. P. von Schroeder
University of Toronto, Toronto Western Hospital, Toronto, Canada

B. Weyand
Hannover Medical School, Department of Plastic, Hand and Reconstructive Surgery, Hannover, Germany

J. Kramer
Newton, MA 02468, USA

S. Bohlmann · M. Reuter (✉)
Technical University, Clausthal-Zellerfeld, Germany
e-mail: matthias.reuter@tu-clausthal.de

M. Israelowitz et al. (eds.), *Biomimetics and Bionic Applications with Clinical Applications*, Series in BioEngineering,
https://doi.org/10.1007/978-3-319-53214-1_12

processing large data systems [3]. *Neural networks* can be considered as descriptive modeling much like more traditional analytic tools, but *neural networks* can reduce the need for computer power for every parameter that is introduced by analytic models [4]. In this way they can be used for undertaking enormous challenges such as artificial modeling o of the brain [5, 6].

Metabolic pathways can be considered as a series of enzyme-catalyzed reactions where the product of one reaction becomes the substrate for subsequent reactions and where all reactions are branched and interconnected as a graph network. Metabolic pathways have traditionally been modeled by using Michaelis-Menten kinetics [7]. Although the equation's stable predictions work well in a homogenous system, it becomes very limited when the model is complex, and especially for reverse reactions, and hence the Michaelis-Menten kinetics does not necessarily describe enzyme biology accurately. To overcome this limitation, particularly for complex system interactions in metabolic pathways [8], we propose to use neural networks via Cell Communication Protocol© by using:

1. *Self-Organizing Map* (*SOM*) *or Self-Organizing Feature Map* (*SOFM*): a type of artificial *neural network* that is trained using unsupervised learning to produce a low-dimensional (typically two-dimensional), discretized representation of the input space of the training samples, and called a map. Self-organizing maps are different than other artificial neural networks in the sense that they use a neighborhood function to preserve the topological properties of the input space [9, 10].
2. *Neural Gas*: The neural gas is a biologically inspired adaptive algorithm. It sorts for the input signal according to how far away they are. A certain number of them are selected by distance in order, and then the number of adaptation units and strength are decreased according to a fixed schedule. *Neural gas* is a special kind of SOM, but here the fixed neighborhood of the SOMs does not exist. As such, even high dimensioned feature spaces can be analyzed or clustered; this could only be done by SOMs in special cases as they are 2 dimensioned [11].
3. *Back Propagation* (*BPP*): Propagation of error is a common method of teaching artificial *neural networks* how to perform a given task. It is a supervised learning method, and is an implementation of the delta rule. It requires a teacher that knows, or can calculate, the desired output for any given input. It is most useful for feed-forward networks (networks that have no feedback, or simply, that have no connections that loop). Back propagation requires that the activation function used by the artificial neurons (or "nodes") is differentiable [12, 13].

The model can be tested against Kyoto Encyclopaedia of Genes and Genomes (KEGG) [14, 15], BioCyc database collection, MetaboLights, Molecular Ancestry Network (MANET) [16, 17] and Reactome [18].

2 Model

The *neural network* model (Fig. 1a) can be applied to metabolic pathways [19], and diffusion models. We considered three approaches to maximize the effectiveness of prediction by the neural network via Cell Communication Protocol.

The first approach was by training the self-organizing map (SOM) by using competitive learning. When a training example is fed into the network, its Euclidian distance to all weight vectors is computed. The '*neuron*', whose weight is most similar to the best matching input (BMU) and the neuron closest to the SOM lattice, are adjusted towards the input vector; the update is given by [20]:

$$Wv(s+1) = Wv(s) + \theta(u, v, s)\alpha(s)(D(i) - Wv(s)) \tag{1}$$

where, s is the step index, i index for the training set, u is the weight for BMU, for $D(i)$, $\theta(u, v, s)$ is the neighborhood function depending on the lattice, $\alpha(t)$ is a monotonically learning function, $D(i)$ is the input vector, v is all the visit vectors. The SOM option shows the developed network [21].

The second approach uses neural gas which considers a probability distribution, $P(i)$ of data vectors i. Every time the s step is a random choice, with the V_j, $j = 1, \ldots, N$:

$$V_{j_k}^{s+1} = V_{j_k}^{s} + \varepsilon e^{-k/\lambda}\left(i - V_{j_k}^{s}\right) \tag{2}$$

The second approach is useful to limit the pathway distances, *i.e.* it tells us the limitations of the metabolic pathway [22, 23].

The third approach is back propagation based on the delta rule (Fig. 1b) where the initial weight is random. During the training, the weights are obtained from the input layers.

The input is a set of tuplets $(i_1, i_2, \ldots, i_n)$, and (t) corrects the output. Because the initial weights are random, the weight from the training set is:$y = i_1w_1 + i_2w_2$ (*NB* because back propagation treats each neuron in isolation, gradient calculation considers only one input at a time); therefore the square error is $E = \frac{1}{2}(t - y)^2$[24].

The neuron activation is given by

$$a = \sum_j i_j * w_j \quad \text{(Figure1c)} \tag{3}$$

The third approach allows handling of reverse catalysis which is one of the limitations of the Michaelis-Menten Kinetics [25].

Fig. 1 **a** Neural network modelling allows inputs with different weights to be processed differently (via a sigmoid function). **b** Neural network modelling can allow for multiple connections to give a solution (output). **c** Detailed diagram of different weights reduced to a sigmoid function to give output activity

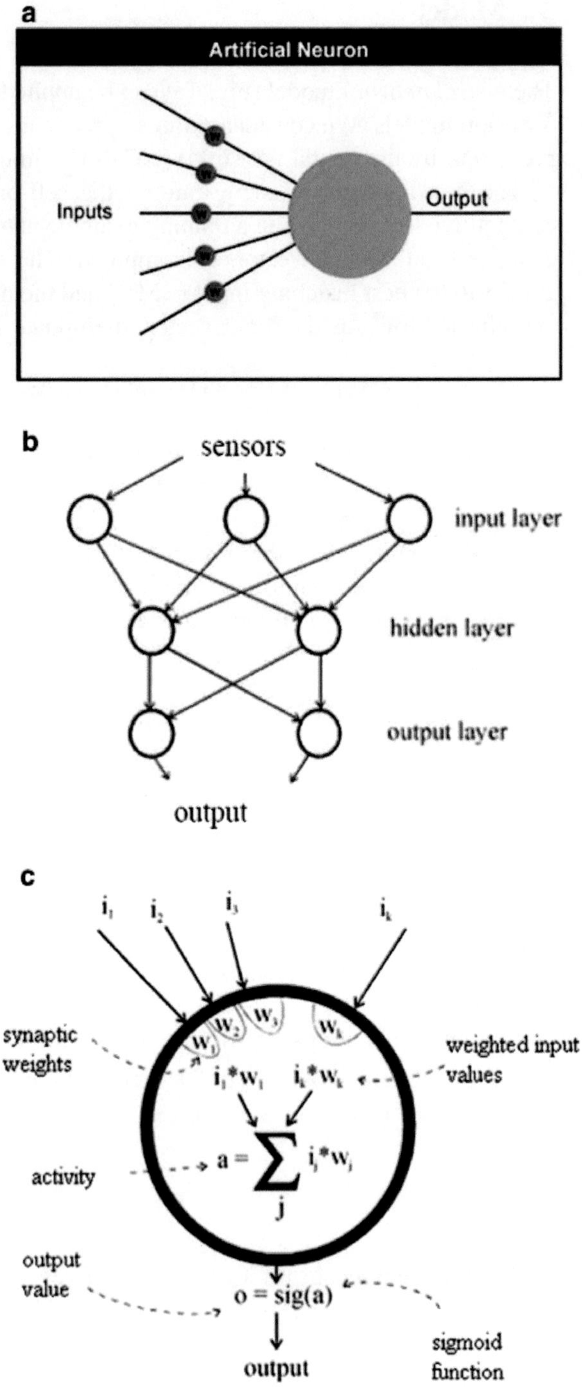

3 Method

Metabolic pathways are a series of reactions where one reaction produces the substrate for the subsequent reactions; neural networks can match the model. The initial random weight entry incorporates the constant parameters and particle density from an enzyme assay. Specifically, the input vector is given by weight obtained from density of the enzyme assay. To develop the chemical pathway the number of clusters representing an assumed network for the metabolic pathway, and a maximal range of clusters nodes and possible nodes for the pathway are entered. Figure 2, shows input vectors for the training set that sets up the actual training of the network.

4 Results

The diffusion across the training set is shown in Fig. 3a, where the upper cells are the initial vectors from the weight given by the density of the enzyme. The reference vector is the upper part; in this example the simulation considered 100 input neurons. The input vector is obtained by the weight of the close nodes (neurons) to the SOM lattices with adjustment [26].

Figure 3c shows the activity card, where the lattices are followed in real time. The second screen is the development of the network and shows the data point of the nodes which are also evolving in time: the red points are clusters or nodes representing products of enzyme catalysis, and the green lights are the references neurons where the pathway is being developed. The yellow is the reference input vector. In Cell Communication Protocol there is an option for changing the positions to alter the graph, the direction and the catalysis, all in real time.

Figure 3d shows the data representing the evolving network in two-dimensions; each cell is represented by the intensity of the color of the nodes, clusters and

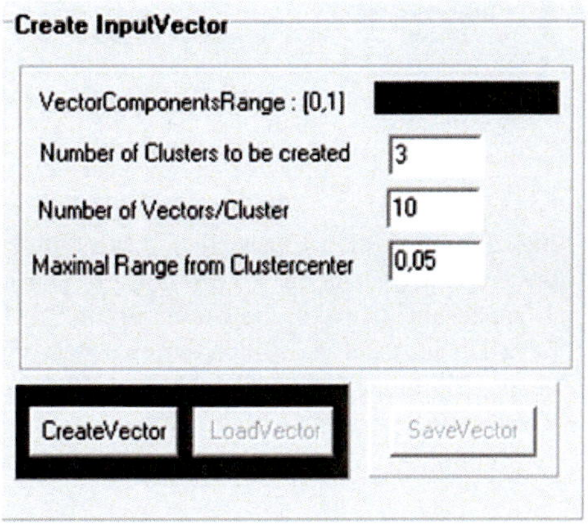

Fig. 2 Computer screen for input vector, which represent the input space of the training sample, the data come from the database being analysed, and the data is defined as input vector (SOM Modus)

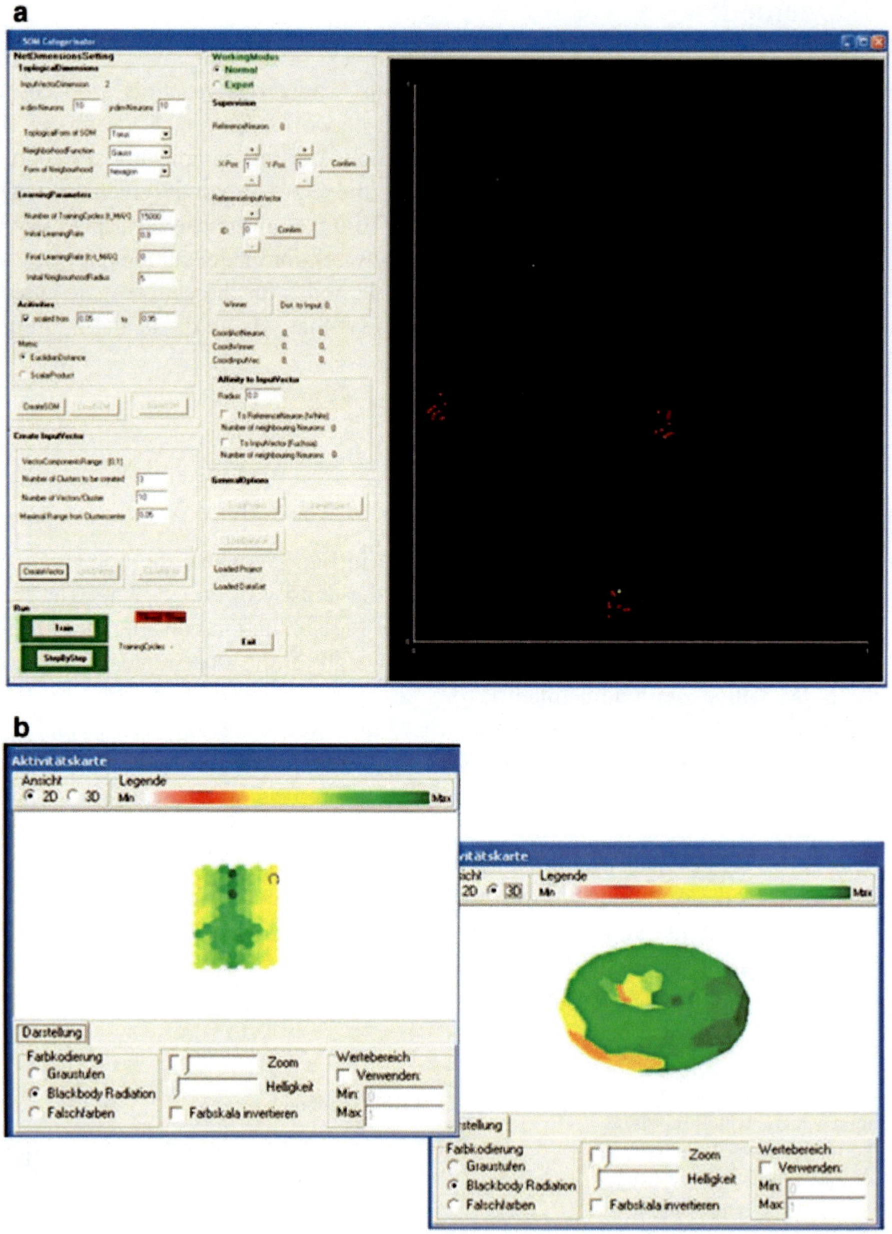

Fig. 3 **a** The cluster consisted of vectors show in red, the green are the input vectors, the yellow is the reference vector (SOM Modus). **b** The activity card of the Neuron in 2-Dimensions and 3-Dimensions (show in the Torus) in the SOM Modus. **c** The number of Neurons and the option for 3-Dimensional and 2-Dimensional data visualization in the Neural Gas Modus. **d** The data being train in the Neural Gas Modus and the data visualization of the training as 3-Dimensional visualization via torus representation

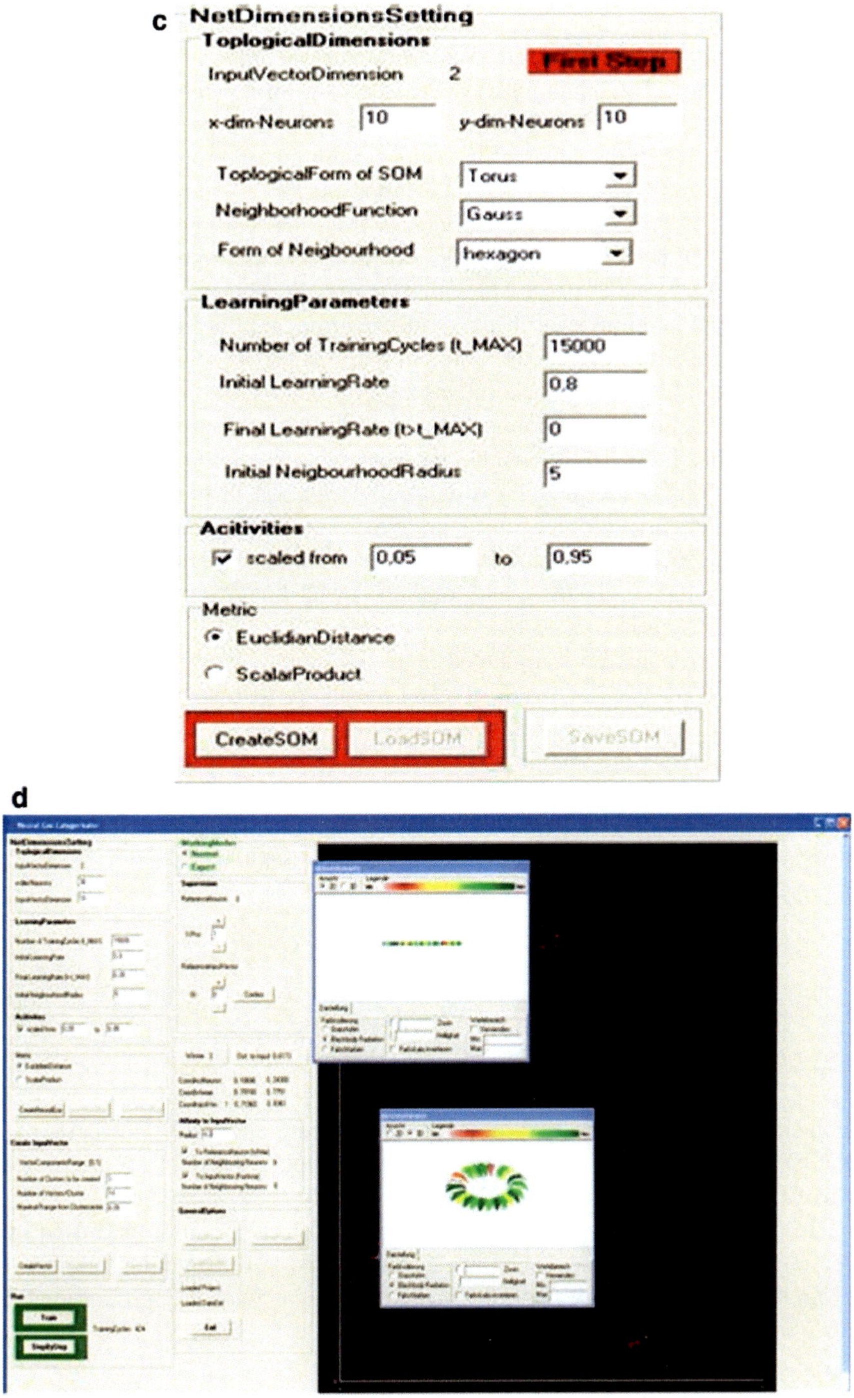

Fig. 3 (continued)

chemical reactions in the development of the metabolic pathway. In three-dimensions, the increased data improves the rate of accuracy. The 3-D visualization in Cell Communication Protocol is a torus, and the changes of color during the simulation gives a better handle of the data points of the metabolic pathway [27]. When the edges of the cells are joining, the topology is of the torus.

For the neural gas approach, an input vector is introduced to the Cell Communication Protocol, and similar to SOMs, the learning parameters are calculated from the input vector (see Fig. 3a) [28]. The input vector parameter is the density of the enzyme as per the previous version. SOMs are under constraints [29] of a predefined neighborhood between the neurons and the distances between the vectors that are not directly represented in the map [9, 30]. The output space is not defined in neural gas and the neural gas algorithm is based on a stochastic gradient decent cost function and is connected to a Hebbian competitive rule which forms the connections between the neighborhoods [31].

Figure 3b shows the neural gas data representation in two-dimensions. At the base, the neural gas prototype clusters move around in the data space similar to the Brownian movement of gas molecule in a closed container. The visualization shows the data structures, clusters, and outliers in 2-dimensions as a line and in 3-dimensions as an iris [32–34].

The back propagation is included in Cell Communication Protocol© to better mimic reverse catalysis of real biological metabolic pathways which are not considered by Michaelis-Menten Kinetics [35, 36].

Figure 4a shows the learning rule, which considers the desired output for any given input, and the momentum term which is a component that dampens the oscillations of the learning rule to provide direction. Figure 4b shows the overall error, where the rate of learning is given by the probability (y-axis) and the interactions in the x-axis.

5 Discussion

The particle diffusion (i.e. mobility) impacts the reaction of the enzyme kinetics. Collision theory gives the enzyme catalysis reaction, and the bimolecular reaction can be expressed as collision [37]. The original Stokes-Einstein diffusion equation was developed for small particles, but can also be used to predict for sizes up to micrometer [38–40]. The diffusion equation is a partial equation and used in concentrations where viscous forces are dominated by low Reynolds numbers [41]. From the low Reynolds number, the particle moves at a velocity proportional to the force applied:

$$v_s = \sigma F \tag{4}$$

where, σ is the proportionality constant ($\sigma \propto 1/r\eta$).

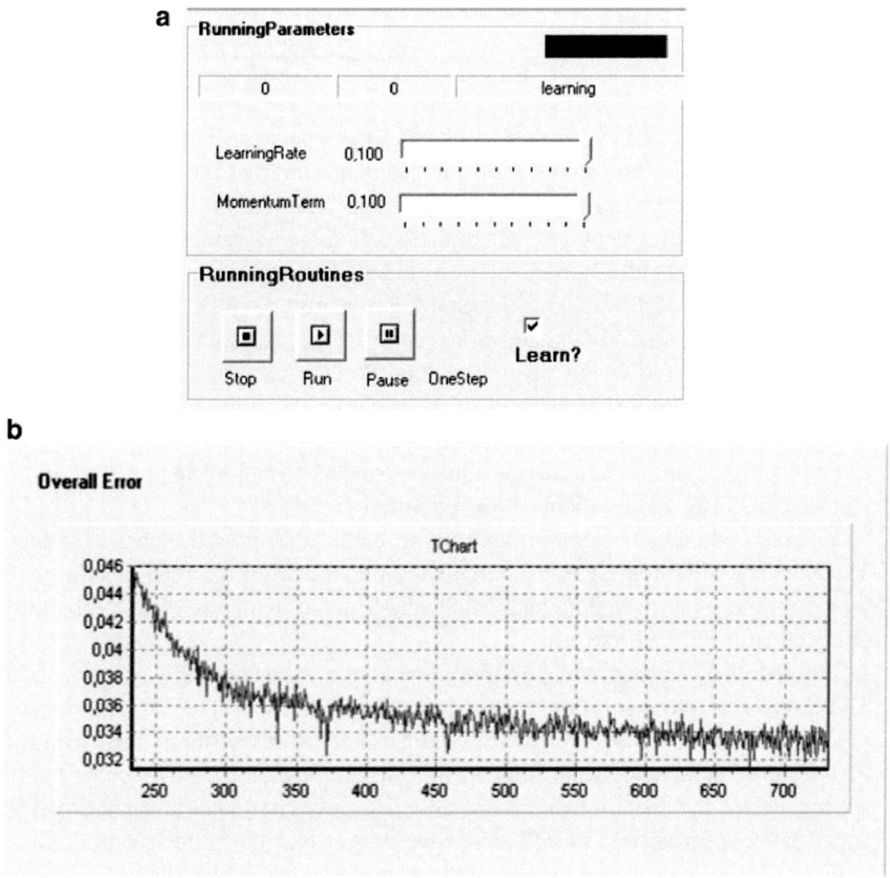

Fig. 4 **a** The classification for back propagation in process of learn in the data base. **b** The error versus the interactions and the error having a solution in the back propagation modus

$$v_s = \sigma \frac{\mathrm{Re}}{\rho} \tag{5}$$

where, Re is the Reynolds number, η is the dynamical viscosity, ρ is density, and r is the radius of the particle.

The diffusion coefficient is given by:

$$D = \frac{k_B T}{\sigma} \tag{6}$$

where k_B is the Boltzmann constant and T is temperature.

From collision theory we can be relate to Michaelis-Menten Kineticş.

The limitation is given by the Michaelis-Menten kinetics:

$$v = \frac{d[p]}{dt} = \frac{V_{\max}[S]}{K_m + [S]} = \frac{D[S]}{K_m + [S]} \tag{7}$$

where $V_{\max}$ is the maximum rate archive by the system, S is concentration of the substrate and K_m is the Michael constant, which is half of the $V_{\max}$.

The initial weights (*e.g.* by Michaelis-Menten kinetics) for the SOM, produces a graph for a particular network, where each cell shows a catalyst reaction, Fig. 5a.

Figure 5b shows the winner neuron and its distance (i.e. the initial catalysis reaction), and also shows how the SOM has learnt, as well as how far the last reaction is from the initial catalysis reaction [42]. Figure 5c shows the actual reaction with the red being the initial input vector and the white marks are the number of the neighboring neurons to the reference neuron and their arrangement.

Figure 5d is the same representation for neural gas, where X-Pos are the reference neuron, and where the weight is from Michaelis-Menten kinetics. The winner cells took 13 iterations from the initial catalysis.

Figure 5e shows the iteration and the probability error of the back propagation. The Michaelis-Menten Kinetics weights are derived from the relationship between the Probability of diffusion and Michaelis-Menten Kinetics. The probability is obtained from Ficks First Law.

For proofing the theoretical assumptions above we used pathway data sets from The Kyoto Encyclopaedias of Genes and Genomes (KEGG) [43, 44] and trained the SOM with these data. We selected two different features of the pathway data. The first feature vector coded the number and distances of the pathways [45] and the second vector the metabolism characterized by the chemical substances involved and their concentrations. In both cases we expect that the trained neural classifier would be able to separate/identify different path-ways constellations if data sets from different experiments were presented to the neural net [46]. The theory of the neural nets (and therefore for every kind of SOM) every constellation of measured data can be combined (used) to form an input or training vector [47].

Figure 6a shows an untrained trained SOM with 100 neurons (green crosses) in a 2 dimensional feature spaced which was normalized in an [0, 1] interval. This normalization is used very often, to prevent numerical overflows in the calculation during the training phase of the SOM. The shown 10 red point in Fig. 6a represent the constellation of a pathway metabolism.

After 15,000 training iterations (requiring about 4 min of computational time) the green neurons adapted to the model noted by localization of the green neurons near the red *pathway model feature points* as shown in Fig. 6b.

In Fig. 6c similar data sets from other measurements has been presented to the net. As can be pointed out, only one of them met the classifiers structure (the yellow neuron in the middle of Fig. 6c), the other data sets—represented by the 2 white and the 10 purple neurons—present similar but not the same structure as the trained ones.

One of the major advantages of neural nets as be able to chosen method allows defining (e.g. out of theoretical consideration) an arbitrary input vector which can

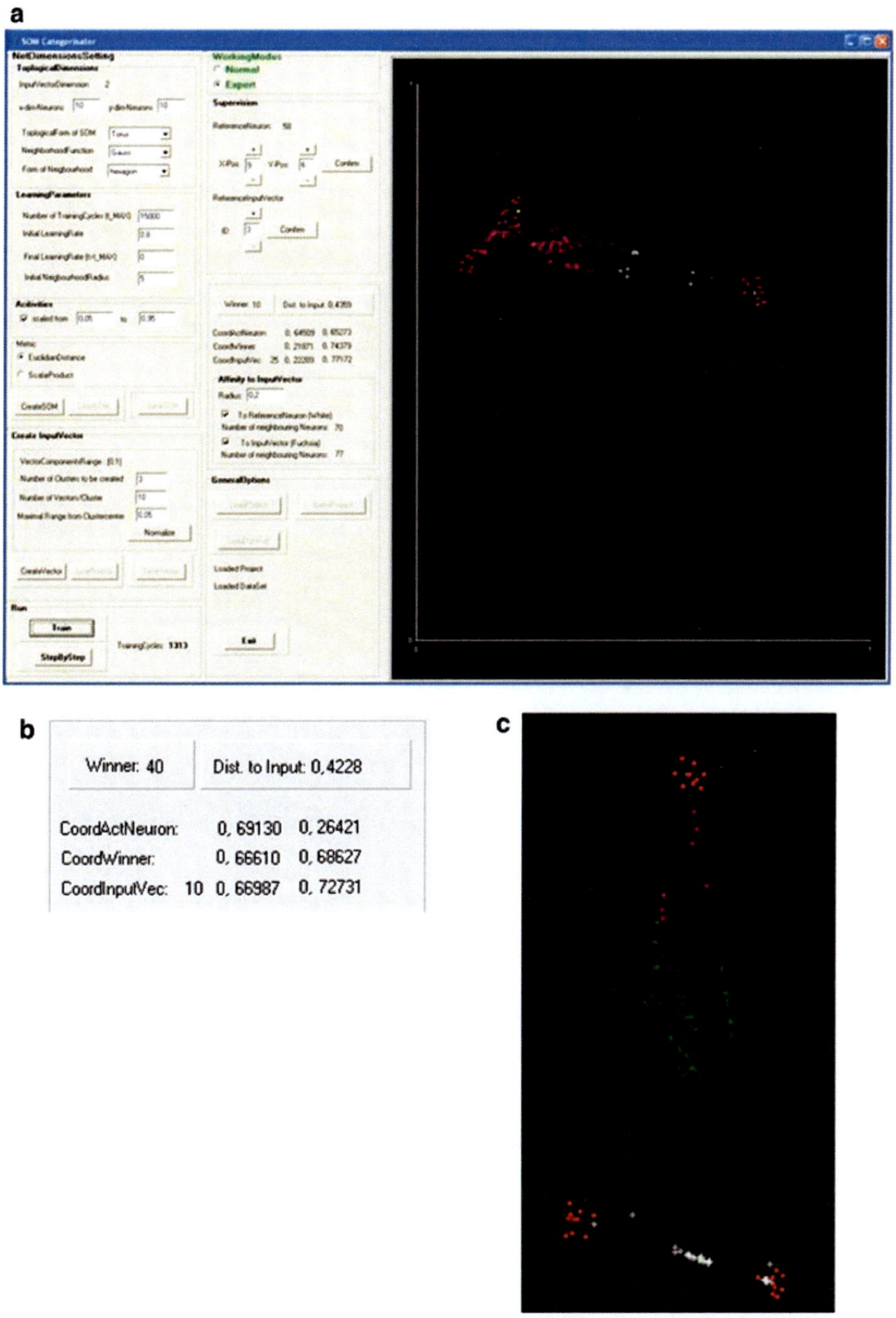

Fig. 5 **a** Training set of the data bases in the SOM modus, the yellow point shows the best solution and red, model feature points. **b** The interaction that has a solution in the data base in the SOM modus, given the distances to solution at 0.4228, the winner took 49 interaction and the location of the best solution. **c** The output of the data base in the SOM modus, and solution is to be finding in the upper part of the simulation. **d** The Neural Gas Modus in a 2-Dimesion data representation with 13 interactions to obtain the best solution, with coordinates locating the solution. **e** The back propagation output neuron with a solution of the data base, the x coordinated shows the interaction and the y axis the error approximation

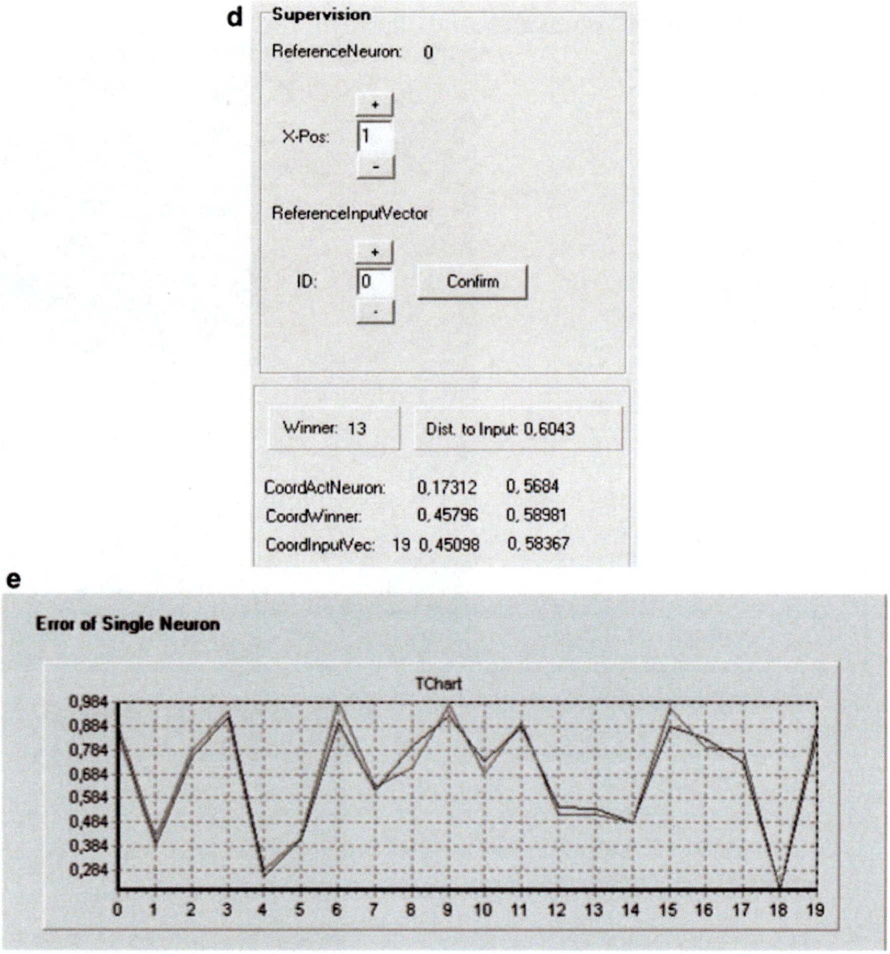

Fig. 5 (continued)

be validated by SOM regarding its fitness [48]. Occasionally, especially if the feature vector and/or the model was more or less complicated, it was be necessary to enlarge the number of neurons of the SOM and/or the number of training cycles [49]. This case is shown in Fig. 6d and e.

Fig. 6 **a** Structure of a so far untrained SOM, of the data base distances between the data is ▶ between 0 and 1 and the best solution is the yellow point, the green points are data distribution and the red point's model feature points. **b** Trained SOM, shows smooth distribution of the data in green, with the red points are model feature points. **c** Evaluation of similarity/identical structure of variable data set by a trained SOM. **d** Not sufficient trained SOM, the distribution of data and the data is not smooth, the red point, are feature points with irregular distribution. **e** Sufficient trained SOM, the green points a smooth data and the red point shows minimize model feature points. **f** Common and modified visualisation of the activity of a trained SOM, the two Dimensions data representations shows, the data distribution and the smoothness of the data

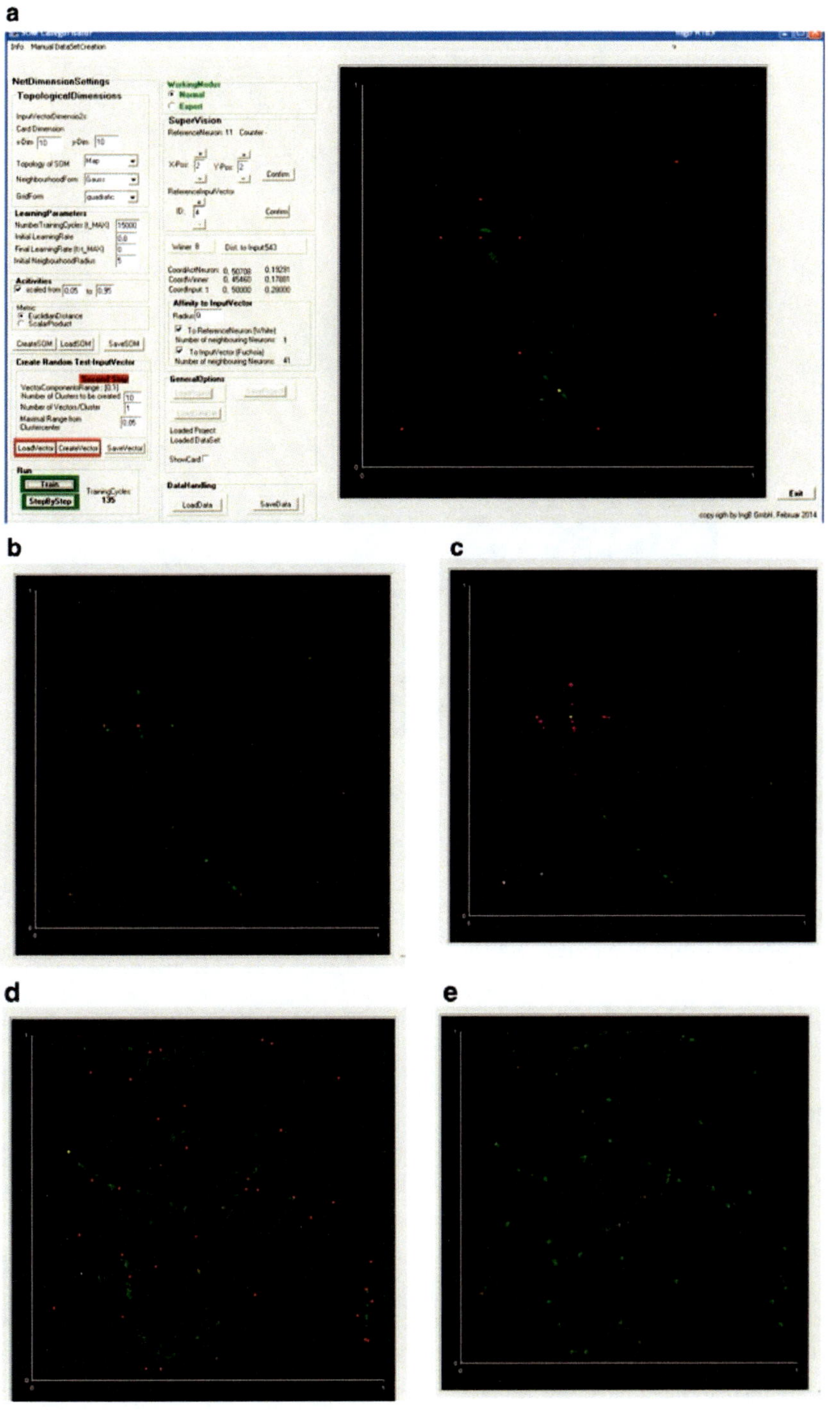
a
NetDimensionSettings
TopologicalDimensions
LearningParameters
WorkingModus
Normal
Expert
SuperVision
Affinity to InputVector
Create Random Test InputVector
GeneralOptions
Run
Train
StepByStep
DataHandling
LoadData
SaveData
Exit
b
c
d
e

f

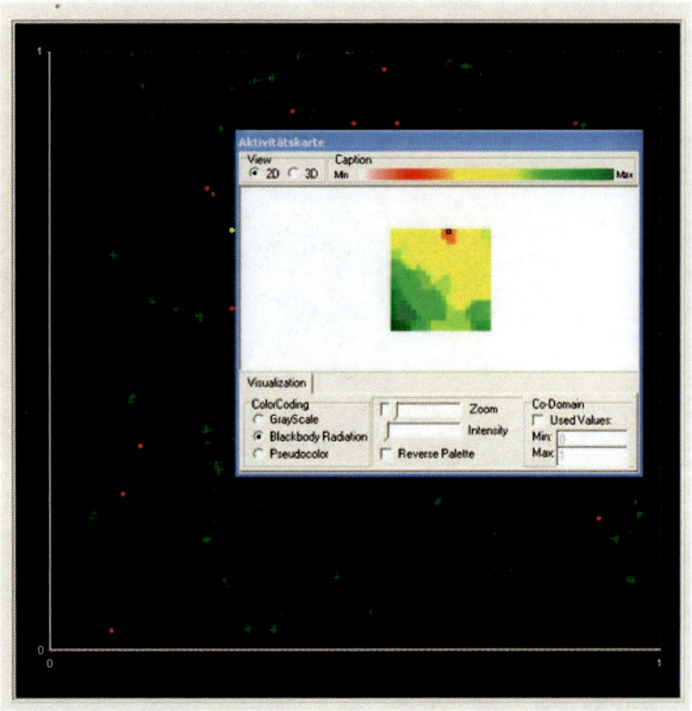

Fig. 6 (continued)

Figure 6d shows—due to the fact that now 44 features (red points (model feature points) in Fig. 6d and e) represent the model—the SOM based classifier now bears 400 neurons, but even after 5.426 training cycles the net structure has not adapted the red points (the system forming the classifier).

It took 7,400 training cycles more until the net structure fits the system structure, as shown in Fig. 6e.

Neural nets should not be relied upon for automatic supervision and deciding [50] if a net structure fits or if the model or system will adapt. But, this model allows a visualisation routine to aid the user to find and define a satisfactory decision if a neural net met a desired classification or not [51].

Figure 6f shows the common visualisation of a SOM activity. Here the so called winner neuron (the yellow neuron, left hand side to the SOM activity card) is represented in orange, limited by a black square at the top. The activity of the other neuron—from light orange over yellow to dark green are surrounding the winner, but—and this is most important for our discussing—no information about similarity or identity can by pointed out by this visualisation. So it is the special way of visualisation we discuss here, which enables a deeper model analysis and/or competition of measured data set.

6 Conclusion

Metabolic pathways are difficult simulated with Michaelis-Menten kinetics that do not consider the reverse catalysis. By using a neural network, Self-Organizing Maps, and back propagation, this methodology can overcome the Michaelis-Menten limitations. Self-Organizing Maps are used to graphically describe the catalysis reaction, with the constraint of the weight space being overcome by the neural gas algorithms [52–56], and by doing so achieves a visual display of the catalysis and expands the weight space to improve the accuracy of the modeled reaction [57–60]. With the use of the back propagation the model can overcome the inherent limitation of Michaelis-Menten, and the Cell Communication Protocol© provides a set of solutions for a descriptive and analytical analysis of metabolic pathways.

References

1. Israelowitz, M., Rizvi, S.W., Kramer, J., von Schroeder H.P.: Computational modeling of type I collagen fibers to, determine the extracellular matrix structure of connective tissues. Protein Eng. Des. Sel. **18**, 329–335 (2007)
2. Israelowitz, M., Weyand, B., Rizvi, S.W., Gille, C., von Schroeder, H.P.: Protein Modeling and Surface Folding by Limiting the Degree of Freedom. Springer, New York Dordrecht London (2013)
3. Sole, R., Delgado, J.: Universal computation in fluid neural networks. Complexity **2**, 49—55 (1997)
4. Bahiraei, M., Hosseinalipur, S., Zabihi, K., Taheran, E.: Using neural network for determination of viscosity in water-TiO2 nanofluid. Adv. Mech. Eng. (2012)
5. Reuter, M., Lenk, K., Schroeder, O., Gramowski, A., Jügelt, K., Priwitzer, B.: Information extraction from biphasic concentration-response curves for data obtained from neuronal activity of networks cultivated on multielectrode-array-neurochips. Presented at the BMC Neuroscience January (2010)
6. Lodhi, H., Muggleton, S.: Modeling metabolic pathways using stochastic logic Programs-Base Ensemble methods. Comput. Methods Syst. Biol. 119–133 (2005)
7. Duggery, R.G., Clarke, R.B.: Experimental design for estimating of the Michaelis-Menten equation from progress curves of enzyme-catalyzed reactions. Biochim. Biophys. Acta. **1080**, 231—236 (1991)
8. Keener, J., Sneyd, S.: Mathematical Physiology I: Cellular Biology. Springer
9. Kohonen, T.: Self-organized formation of topologically corrected feature map. Biol. Cybern. **43**, 59–69 (1986)
10. Reuter, M.: Of the Stability of Closed Self Organising Maps (gSOMs) for Predictive Control. Presented at the, Lyon, France (2008)
11. Martinez, T., Berkovich, S., Schulten, K.: Neural Gas Network for Vector quantitation and its application to time series prediction. IEEE-Trans. Neural Netw. **4**, 558–569 (1993)
12. Rumelhard, D.E., Hinton, G.E., William, R.: Learning representations by back propagation errors. Nature **323**, 533–536
13. Gorban, A.N., Zinovyed, A.: Principal Manifolds and graphs in practice: from molecular biology to dynamical systems. Int. J. Neural Syst. **20**, 219–232 (2010)

14. Kanehisa, M., Goto, S., Kawashima, S., Okuno, Y., Hattori, M.: The KEGG resource for deciphering the genome. Nucleic Acids Res. 32(Database issue), D277–80 (2004)
15. Kanehisa, M., Goto, S., Hattori, M., Aoki-Kinoshita, K.F., Itoh, M., Kawashima, M.: From genomics to chemical genomics: new developments in KEGG. Nucleic Acids Res. **34** (Database issue), D354–7
16. Salek, R.M., Haug, K., Conesa, P., Hastings, J., Williams, M., Mahendraker, T., Maguire, E., González-Beltrán, A.N., Rocca-Serra, P., Sansone, S.A., Steinbeck, C.: The MetaboLights repository: curation challenges in metabolomics, 1. Database. 2013, bat029 (2013)
17. Kim, H.S., Mittenthal, J.E., Caetano-Anolles, G.: MANET:tracing evolution of protein architecture in metabolic networks. BMC Bioinf. **7**
18. Croft, D., O'Kelly, G., Wu, G., Haw, R., Gillespie, M., Matthews, L., Caudy, M., Garapati, P., Gopinath, G., Jassal, B., Jupe, P., Kalatskaya, I., Mahajan, S., May, B., Ndegwa, N., Schmidt, E., Shamovsky, V., Yung, V., Birney, E., Hermjakob,, H., d' Eustachio, P., Stein, L.: Reactome: A database of reactions, pathways and biological processes. Nucleic Acids Res. 39 (Database issue), D691–D697
19. Gorban, A.N.: Principal Manifold for Data Visualization and Dimension Reduction. Springer
20. Shah-Hosseni, H., Safabakhsh, R.: TASOM: A new time adaptive self-organization map. IEEE Trans. Syst. Man Cybern.-Part B Cybern. 271–282 (2003)
21. Reuter, M., Bohlmann, S.: Supervising MultiCut Aggregates by Special Neural Nets. Presented at the WAC 2012, Puerto Vallarta, Mexico (2012)
22. Canales, F., Chacon, M.: Modification of the growing neural gas algorithm for cluster analysis. Presented at the International Association for Pattern Recognition, Image Analysis and Applications 12th Iberoamerican Congress on Pattern Recognition, CIARP 2007, Viña del Mar-Valparaiso, Chile, November 113–17 (2007)
23. Reuter, M.: Supervising cathodic protected gas nets with CI-based methods. Presented at the ISC'2013, 11th Annual Industrial Simulation Conference, Ghent, Belgium, 22 May 2013
24. Rojas, R.: Neural Networks- A Systematic Introduction. Springer-Verlag, Berlin, New York (1996)
25. Goudar, C.T., Sonnad, J.R., Duggley, J.R.: Parameter estimation using a direct solution of integrated Michaelis-Menten equation. Biochim. Biophys. Acta- Protein Struct. Mol. Enzymol. 1424, 377–383
26. Kagans, I.: Time-dependent self-organization maps, (1994)
27. Kurasova, O., Molytė, A.: Quality of quantization and visualization of vectors obtained by neural ga and self-organizing map by neural ga and self-organizing map. Informatica **22**, 115–134 (2011)
28. Estėvez, P.A., Figueroa, C.J.: Online data visualization using neural gas network. Neural Netw. 19, 923–934
29. Kolbe, L., Tünnerman, R., Hermman, T.: Growing neural gas sonification model for interactive surfaces. In: Proceeding ISon 2010 3rd Interactive Sonification, KTH, Stockholm, Sweden (2010)
30. Kaski, S.: Data Exploration using self-organizing maps. Acta Polythecnica Scand. Math. Comput. Manag. Publ. Finn. Acad. Technol. 57–60 (1997)
31. Martinez, T.M., Martinez, K.J.: Topology representing networks. Neural Netw. **7**, 507–522
32. Liu, K., Liu, P.: Color model based 3-D self-organizing map, information visualization. Inf. Vis. IV pp. 403–408 (2004)
33. Xinzhi, L.J.: Visualization of high-dimensional data with relational perspective map. Inf. Vis. **3**, 49–59 (2004)
34. Ito, M., Myoshi, M.: The characteristic of torus self-organizing map. Presented at the In Proceeding 16th Fuzzy Systems Symposium Akita, Japan Society for Fuzzy and Systems, Japan (2000)
35. Riedmiller, M., Braum, H.: A direct adaptive method for faster backpropagation learning: The RPROP Algorithm, Neural Networks. Presented at the IEEE International Conference (1993)
36. Riedmiller, M.: Advanced surpervised learning in multi-layer perceptrons—from backpropagation to adaptive learning algorithms. Comput. Stand. Interfaces. **16**, 265–278

37. Blanch, B.H., Clark, D.S.: Biochemical Engineering. Marcel Dekker, New York (1997)
38. Bremmell, K.E., Wissenden, N., Dustan, D.E.: Diffusing probe measurements in newtonian and elastic solutions. Adv. Colloid Interfaces Sci. 89–90, 141–154
39. Norris, D.A., Sinko, P.A.: Effect of size, surface charge, and hydrophobicity on the translocation of polystyrene microspheres through gastrointestinal mucin. Appl. Polym. Sci. **63**, 1481–1492 (1997)
40. Crocker, C.J.: Measurement of hydrodynamic correction to the Brownian motion of two colloid spheres. J. Chem. Phys. 106, 2837–2840 (1997)
41. Weyand, B., Israelowitz, M., von Schroeder, H.P., Vogt, P.: Fluid dynamics in bioreactor design: considerations for the theoretical and practical approach. Bioreactor Systems for Tissue Engineering. pp. 521–268. Springer, Berlin, Heidelberg, New York (2009)
42. Reuter, M.: Computing with Activities V. Experimental proof of the stability of closed self organizing maps (gSOMs) and the potential formulation of neural nets. Presented at the WAC 2008,, Waikoloa, Hawaii, USA (2008)
43. Caetano-Anollés, G.: Database was retrieved from SCOP 1.67, KEGG Enzyme, and Phylogenetic tree of Protein Fold Architecture, (2004)
44. Retrieved September 2014 from http://www.genome.jp/kegg/pathway.html#metabolism
45. Lotz K., Bilonu L., Roska T., Hamori J.: A cellular neural network model of the time coding pathway of sound localization-hyperacuity in time, Neural Networks, 1996., IEEE International Conference on (Volume:2). IEEE (1996)
46. Lampert, C.H., Nikisch, H., Harmeling, S.: Attributed-base classification zero-shot-visual object categorization. IEEE Trans. Pattern Anal. Mach. Intell. **36**, 453–465 (2014)
47. Grunz, A., Memmert, D., Perl, J.: Tactical Pattern Recognition in soccer games by means of self-organizing maps. Human Move. Sci. **31**, 334–343 (2012)
48. Komendantskaya, E.: Unification neural networks: unification by error-correction learning. Logical J. IGPL **19**, 821–847 (2011)
49. Kröger, B.J., Kannampuzha, J., Kaufmann, E.: Associative learning and self-organization as basic principles for simulating speech acquisition, speech production, and speech perception. EPJ Nonlinear Biomed. Phys. (2014). https://doi.org/10.1140/epjnbp15
50. Stein, G., Chen, B., Wu, A.S., Hua, K.A.: Decision tree classifier for network intrusion detection with GA-based feature selection. ACM-SE **43**(2), 136–141 (2005)
51. Twomey, J.M., Smith, A.E.: Performances measures consistency, power for artificial neural networks models. Mathl. Comput. Model. **21**, 243–258 (1995)
52. Reuter, M., Bohlmann, S.: Automatic Detection of Buried Utilities in Georeferenced Multi-Sensor Data with Neural Networks. Presented at the TOK, Izmir, Turkey August (2011)
53. Goodwin, C.R., Sherrod, S.D., Marasco, C.C., Bachmann, B.O., Schramm-Sapyta, N., Wikswo, J.P., McLean, J.A.: Phenotypic mapping of metabolic profiles using self-organizing maps of high-dimensional mass spectrometry data. Anal. Chem. **86**, 6563–6571 (2014)
54. Ueno K., Mineta K., Ito K., Endo T.: Exploring functionally related enzymes using radially distributed properties of active sites around the reacting points of bound ligands. Struct. Biol. (2012). http://www.biomedcentral.com/content/pdf/1472-6807-12-5.pdf
55. Vikas, Chaudhary V., Ahlawat, A.K., Bhatia, R.S.: Growing neural networks using soft competitive learning. Int. J. Comput. Appl. **21**, 1–6 (2011)
56. Bishop, G.M., Svense, M., Williams, K.J.: GTM: The generative topographic mapping.". Neural Comput. **10**, 215–234 (1998)
57. Psichogios, D.A., Ungar, L.H.: A hybrid neural network-first principles approach to process models. AIChE J. **38**, 1499–1511 (1992)
58. Fernandes, F.A.N., Lona, M.F.L.: Neural network applications in polymerization. Braz. J. Chem. Eng processes. (2005). https://doi.org/10.1590/S0104-66322005000300009
59. Khan S., Xekalakis P., Cavazos J., Cintra M.: Using predictive modeling for cross-program design space exploration in multicore systems. In PACT. (2007) http://homepages.inf.ed.ac.uk/mc/Publications/pact07.pdf
60. Hessine, M.B., Saber, S.B.: Accurate fault classifier and locator for EHV transmission lines based on artificial neural networks. Math. Problems Eng. **2014**, 1–19 (2014)

Computer-Based Intelligence Methods Applied for Personalized Management of Diabetes

Matthias Reuter and Sabine Bohlmann

Abstract Type 2 diabetes has become a typical civilization disease in which modern lifestyle changes are recognized risk factors. Currently, diabetes management is one of the important areas of therapeutic and social medical care. In this paper, we present a computer-intelligence based system for diabetes management and patient trend analysis, which is based on data from a set of questionnaires. Questionnaires are proven tools to handle, evaluate and improve the diabetes management of patients and their acceptance of the disease. Our system was developed based on a large range of questionnaires from American and German diabetes patients. We used the data sets to train a new kind of neural net to create personal fingerprints, enable an individual therapeutic support, and guarantee continuous monitoring over time. Our categorizers differ especially between social, educational and ethnic background and are based on a class-oriented four level statistic. The states of patients' factoring differ between psychological, physiological, familiar and social factors. These have very important influence on the management of type 2 diabetes and the success of therapy as well as the acceptance of the disease itself. Due to its structure, the system can be continuously adapted and sensitized by means of new data. In that way, over time, regional influences are incorporated automatically into the system's categorization and behaviour. In this way, our system leads to a supervision and guidance system for patients and the attending physicians and guarantees at least an overall cost reduction.

Keywords Diabetes · Computing With Activities (CWA) · Personal diabetes management

M. Reuter · S. Bohlmann (✉)
Department of Big Data and Technical Information Systems, TU Clausthal - Institute for Informatics/IngB RT&S, Clausthal-Zellerfeld, Germany
e-mail: sabine.bohlmann@tu-clausthal.de

M. Israelowitz et al. (eds.), *Biomimetics and Bionic Applications with Clinical Applications*, Series in BioEngineering,
https://doi.org/10.1007/978-3-319-53214-1_13

1 Introduction

Diabetes (diabetes mellitus) describes a group of metabolic diseases in which a person has high blood glucose, either because insulin production is inadequate, or because the body's cells do not respond properly to insulin, or both. Diabetes can be classified in two forms: Diabetes Type 1 and Diabetes Type 2 [1, 2].

In type 1 Diabetes there is inability to produce insulin (insulin-dependent diabetes, juvenile diabetes, or early-onset diabetes). Usually this type develops before their 40th year, often in early adulthood or teenage years. Approximately 10% of all diabetes cases are type I. In type 2 diabetes, there is either inadequate production of insulin for proper function or the cells in the body do not react to insulin (insulin resistance). Approximately 90% of all cases of diabetes worldwide are type 2. Sometimes Type 2 diabetes symptoms can be controlled by losing weight, following a healthy diet, doing a lot of exercise, with appropriate monitoring of blood glucose levels. However, Type 2 diabetes is typically a progressive disease and patients will probably end up have to syringe insulin [1, 2]. As both forms differ, every diabetes management system has to distinguish both groups generally, economically, and socio-culturally [3, 4]. The degree of complexity required means that Common expert systems or simpler software may be unable to handle required parameters in the context of an individual finger print of patient's behaviour and disease patterns.

In the last few years, Computer-based Intelligence (CI) has enabled scientists and other stakeholders in medicine and industry to categorize or diffuse uncompleted data by self-learning and in a hyperspace acting algorithm. Together with an adequate encoding, this knowledge enables one to create personal fingerprint software for disease management and to handle diverse data in the context of Big Data [5–7]. In line with this, in the last few years, we developed and tested a special software tool for personalized management of diabetes. Based on a large range of questionnaires from USA and Germany, we trained special neural nets to create a personal fingerprint of different patients (and their medical histories), to form a global categorizer for new patients and/or cohorts of patients with different social, educational and ethnic backgrounds. Furthermore, based on a class oriented four level statistic, we can show, that special analysis of different aspects like psychological, physiological, familiar and social estimates leads to a supervision and guidance system for patients and the attending physicians. That these aspects are import end for a 360° view on the patient was shown in many other scientific, medical-orientated researches [8–16]. The system involves the use of an on-line data bank as well as highly dimensioned questionnaires. The patients answer different questions at their convenience and to the degree to which they are empowered to answer or if the momentary status of a patient being analyzed fit a trend of analysis.

2 Methods

2.1 Basic Modules of the System

Our personal diabetes management system consists of four different modules:

- Data Bank Module (DB), storing all questionnaires and the answers to them, personal data, trends, neural net structures and labels for the different country versions.
- Online and Printing Module (OPM), enabling the connection to the World Wide Web, the email sending/receipt, the report generation and the communication to the printer equipment of the physician's office.
- The Graphical User Interface Software (GUI), enabling the Man-Machine-Interaction (Patient to Computer Dialog, PCD), the help functions and the choice of the language based labeling of the system.
- The CI-Module, involving the neural net structures, the special categorization algorithm, the learning procedures for the system and the evaluation and visualization modules for testing on confidence.

The system itself has four working Modes:

- Patients Mode (PM)
- Doctors Mode (DM)
- Maintenance Mode (MM)
- Condition Mode (CM).

These modes are protected separately by special Personal Identification Number (PIN)-Codes for data security.

2.2 Basic Functionalities of the System

To ensure data security, for every patient and the doctor's office information, there is a 6-letter password that enables the different users to see all personal data, the results of all questionnaires and trend analysis results. Furthermore, the doctor's office function is able to create new patient's "cards", which means new patient profiles stored in the data bank.

If a new patient profile is created, in the first step, all necessary personal data is evaluated by an interactive PCD-dispute, whereby the systems check all data on logic and completeness. In the second step, the patient is lead through the questionnaire. In our system, this questionnaire is divided in 10 logical categories to analyse different items of patient's life and behaviour such as:

- disease status and disease history
- physical status
- psychological status

- social status
- family status
- status of cooperation
- self-assessment
- foreign assessment
- combination of all items listed before for cross-checking.

After the patient has answered all questions, (optional ones in some categories) the system will categorize the patient via its neural structure. These categorization results can be sent to the doctor's office and/or the patient on demand, via internet and/or mobile devices, whereby the results are bundled in a pdf-report.

In the doctor's mode the system can be conditioned, which means special neural nets can be created and trained. Furthermore, the doctor is able to create a statistical and neural overview from all the patient data.

2.3 Principle of the Neural Net Based Categorizer

The core of the system can be found in the neural net structure, which is the underlying data analysis algorithm of a personal diabetes management system. Here we explain some of the basis of these CI-oriented methods. Neural nets are more or less nothing more than a simulation algorithm of central nervous functionalities [17–19]. Therefore, they can be divided in several neurons, which are combined via synaptic connection, whereby these connections can be conditioned in a special training modus. The training itself can be done in two modes, supervised and unsupervised. Supervised training can be done easily if one knows everything about the system to be categorized or examples of all possible states of the system are known. In our case, as we don't know from the beginning how many different classes of patient or (different) states of the disease are existent, this method is not sufficient for a managing system.

The second method to train a network, the unsupervised training, for us, is sufficient as the training algorithm divides out all categories of the data presented to the net during the training phase. However, by common neural nets, there is a problem. At least for every different status of the system/patient, one single neuron has to be implemented or, if we don't know how many states exist, more neurons than existing states have to be implemented in the network.

To work around this challenge, we developed a new kind of processing and interpretation strategy "Computing with Activities" (CWA-Method) [20, 21]. Out of this theory, we interpret the overall activity of the whole net rather than the activity of a single neuron. This interpretation is near to the neurological assumption that the neuron ensemble activity pattern is an accurate model of what happens in the real world, better than classical simulated neural net theory. An example will show us how powerful this new theory is.

Using the classical information theory of the simulated neural nets, we can store in a net, containing 200 neurons, 200 different possible states of a single neuron. If we use the CWA-Method and define that our simulated neurons can occupy five different states, (such as five different firing modes of a fixed time interval of a natural neuron) we can code 5^{200} simulations in the same net. For us, the coding seems to be nearer to nature. This is similar to how a honeybee can communicate information to its colleagues when looking for “flowers”.

Using the CWA-Method also means that slight changes in the data sets will also lead to slight changes in the categorizer result. Within the CWA-Method, the missing link to build a management systems with great differences in patient status will lead to great differences in the activity pattern of our basic categorization module and vice versa.

Figures 1a and 1b show some examples of such an activity based categorizer. The upper examples show a small difference in the data sets, the lower a larger difference. It is important to mention, that the data sets used here have 20

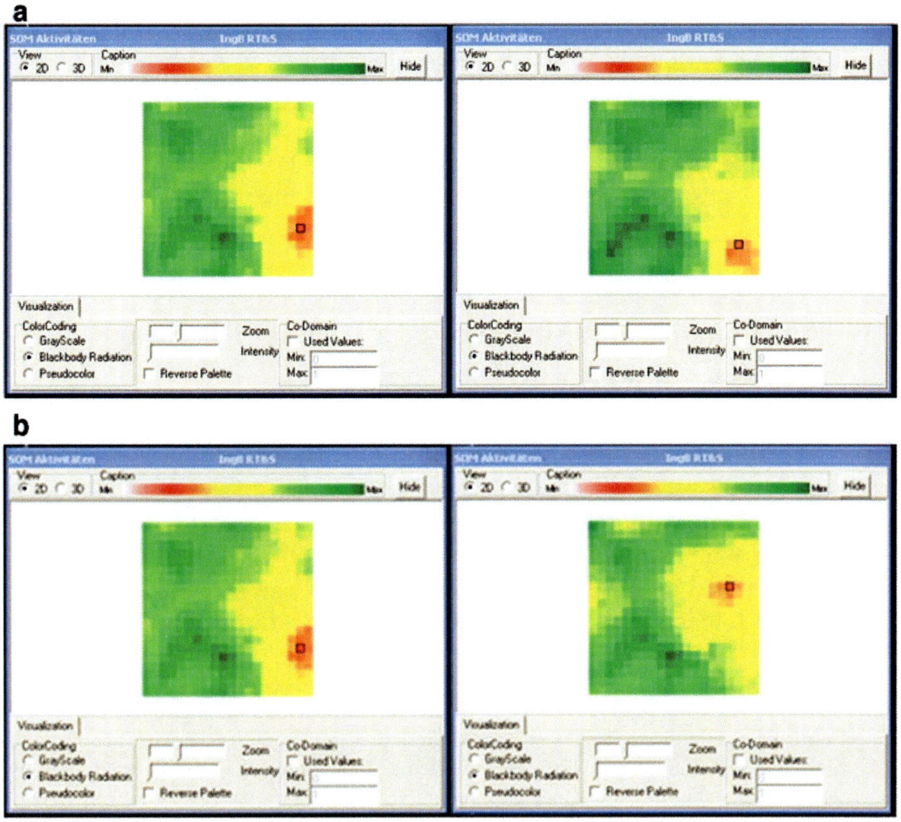

Fig. 1 **a** CWA-Method explained by similar patient’s categorization. **b** CWA-Method explained by unequal patient’s categorization

dimensions, this means that 20 parameters (here 20 different answers to 20 different questions) have been combined to categorize different patients regarding their similarity in behaviour and status.

2.4 Categorizations Criteria, Color Coding and Trend Analysis

Once a patient has answered all questions, the neural net structure categorizes the patient as described above. The system requires an adequate machine-user-interface which enables the medical staff to interpret the neural based categorization in an intuitive and quick way. For this reason, we incorporated a special interpretation module, which analyzes the neural activity pattern in a way that four different general categorization classes describe the momentary state of a patient. These classes are combined with a kind of "next to do list," which presents a reference of therapeutic interventions. These categories and therapeutic interventions are:

- Green: patient's state is OK, no action is needed
- Yellow: patient's state is more or less OK, but has to be supervised and improved
- Orange: patient's sate is not OK, action is needed
- Red: patient's state is alarming, prompt action is needed.

3 Functionality and Structure of a CI-Based Personal Diabetes Management Tool

The realization of a CI-based personal diabetes management tool was taken into account several years ago, whereby we decided to use a C++ based system, connected to a common PostgreSQL Database.

Because of the sensibility of the personal data, all data is encrypted. The system itself is protected by a 9-number PIN code, as shown in Fig. 2.

After logging in as a patient or as a doctor, the patient's personal data is collected by a standard user interface, which is shown by Fig. 3. This electronic form is divided in three main parts:

> Left hand side at the top the "name and address part" and some of the data, like name of birth, first name and date of birth are protected and therefore cannot be changed, even by the medical staff. An extra window represents the employer's part (shown on right hand side). The stored data can be changed and are automatically available for printing a certificate of disability. The actual status part—shown in the left middle part of Fig. 3—represents the actual body of data, like weight, height, gender and the patient actual state; these are data that the system analyzes.

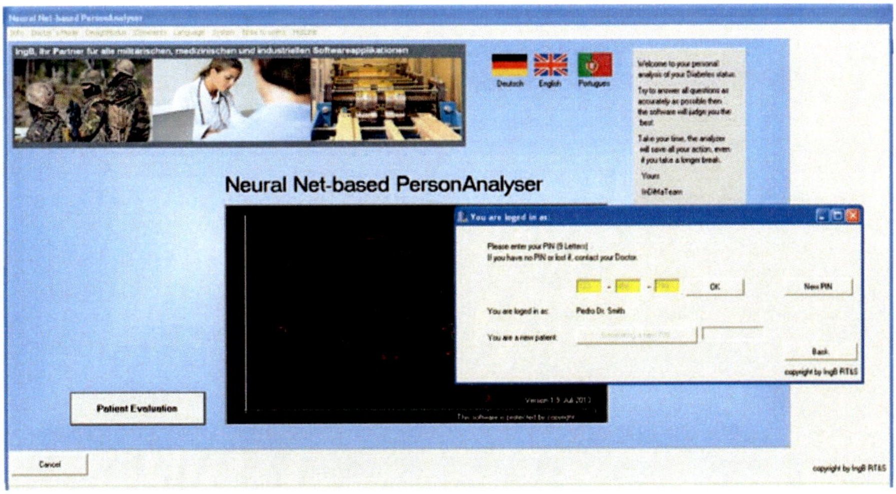

Fig. 2 Start form of the CI-based personal diabetes management tool

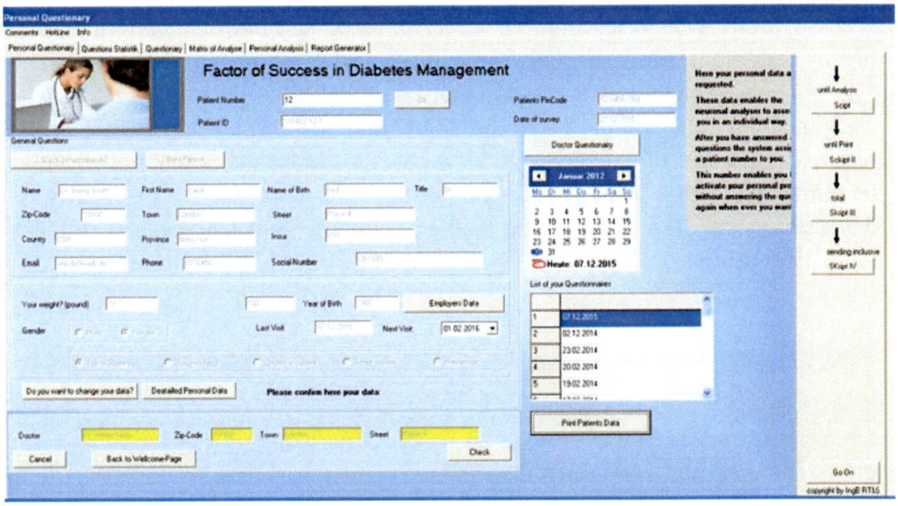

Fig. 3 Patient's data input mask

At the button of the form the doctor's data are shown. The system itself is protected also via this doctor's data; means, changing this data requires a change of the safety keys. At the right hand side of the form some services functions, like a calendar, the history of all sessions of the patient and the printer routines to print the personal file and/or the old reports are documented.

In addition, in this personal data sheet, two other sheets are involved. The first of them involves the data of the employer and the insurance, as shown in Fig. 4.

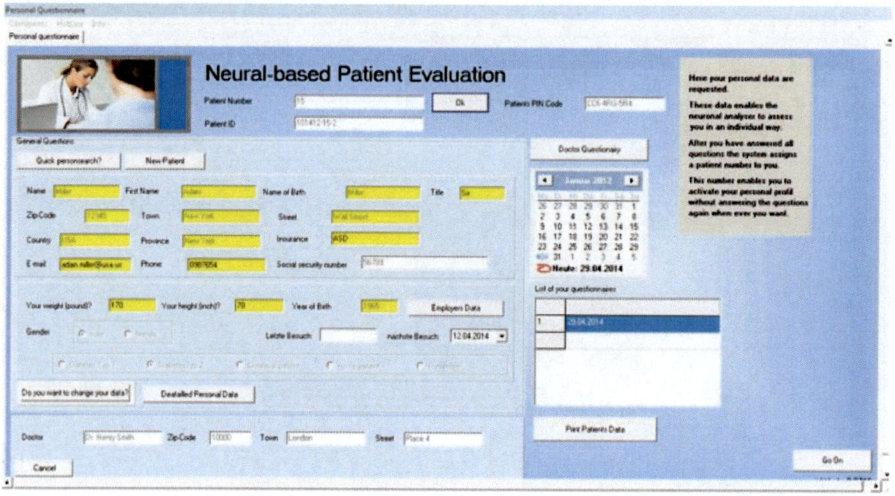

Fig. 4 Patient's data input mask including employer's data

Also, if the "Doctor's Modus" is selected, secondary information of the patient's history, additional diseases and former therapeutic interventions can be recalled. The principal structure of the additional data sheet is visualized in Fig. 5.

Sometimes it is useful to have a short patient's details printed out. For that reason, in the system patient's overview, such a file format is integrated. This can be printed out and/or be included in an existing data collection file or integrated in a database system used in the doctor′s office. Figure 6 shows the general structure of the patient's file.

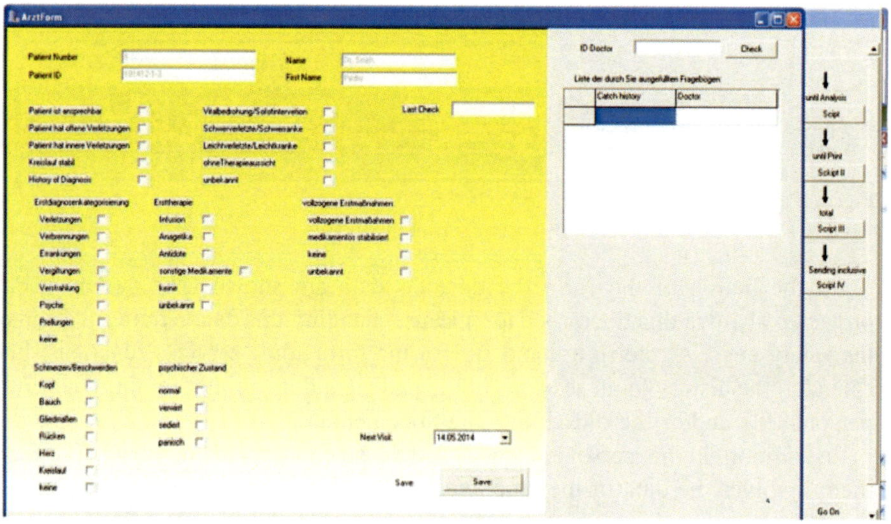

Fig. 5 Secondary patient's data sheet

Ihr Pin:	123-456-789

Name		
	Familienname	Brinkmann
	Vorname	Nora
	Geburtsname	Brinkmann
	Anrede/Titel	Frau
Adresse/Kontakt		
	Land	Deutschland
	Provinz	Niedersachsen
	Postleitzahl	38678
	Stadt	Clausthal
	Strasse	Einersberger Blick 16
	Hausnummer	
	Telefon	05323727104
	Handy	
	Fax	
	Email	xmkdx@web.de
Versicherungsangaben		
	Krankenkasse	dkv
	Versicherungsnummer	21-1234-567-064

Fig. 6 Form and style of the patient's file

After the personal data sheet has been completed, the patient is guided through the different questionnaires under the following categories:

- disease status and disease history
- physical status
- psychological status
- social status
- familiar status
- status of cooperation
- self-assessment, and
- foreign assessment.

Any of these items can be selected from the first sheet of the questionnaire, as shown in Fig. 7.

If an item, also called "key-factor", is completed, the system announces that every question was answered, as shown on the right hand side of Fig. 7. After selecting an item, the system offers the different questions and the possibility to answer via a ranking, as shown in Fig. 8a.

After the questionnaire is completed, the system will ask which items should be analyzed. This is done by a decision mask, as shown in Fig. 8b.

Once the patient and doctor's work inputs are completed, the system generates the diabetes status, the trend and the prognosis of the patient's status with help of the neural nets. All these items are visualized via a simple visual output as shown by Fig. 9.

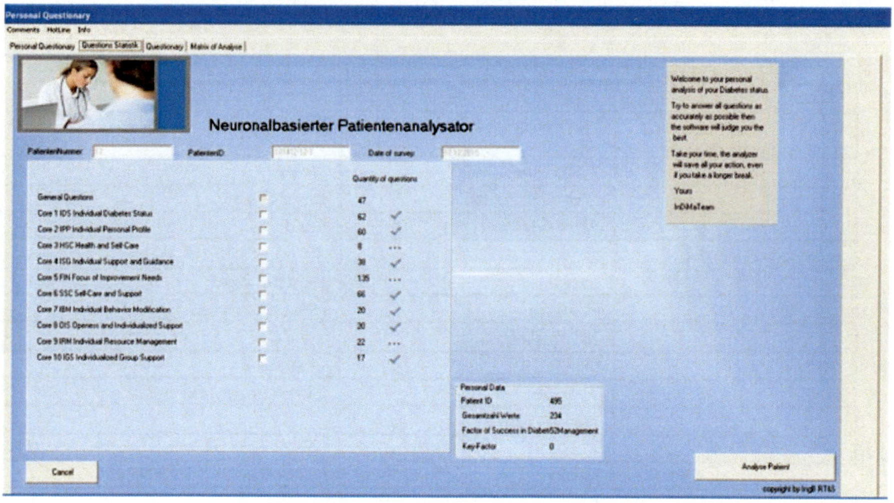

Fig. 7 First sheet of the questionnaire

The color-coded patient state analysis shown in Fig. 9 contain the following:

1. The overall state of the patient, represented by the left hand side block
2. The block in the middle shows the development of the patient in the different categories which are analysed
3. The block right hand side represents the analysis and trend of each question.

These three different categorizations were used to evaluate the system during the R&D-phase of development for best fit of a patient's status. The left columns of each block point out the average of the analysis of all data (here the overall analysis and the trend analysis of the patient's data), whereas, the middle column represents the neural net categorization of the states following the four classes, green, yellow, orange and red. The right column represents the categorization of the neural nets by corresponding rainbow colors. By rainbow colors the status of the patient can be shown much more gradual, because instead of four colors, 256 colors are used. But for a quick overview, it is helpful if a categorization of four colors is presented. This can be pointed out very clearly by the categorization of the single questions (right hand side block). On left hand side, the traditional four color-coding is shown, while the right column shows the categorization by the corresponding rainbow colors. As our investigations showed, the chosen color-coding enables a detailed analysis of patients within "one view", especially the break down from the global categorization of the overall state of the patient leading to an effective estimation of diabetes management.

Figure 9 shows the momentary status of a patient. To ensure that therapeutic measures are successful over the time, or to identify deteriorations on one or several areas, a statistical module which analyses the categorization results over time can be implemented. Based on the database entries of the patients, this module identifies

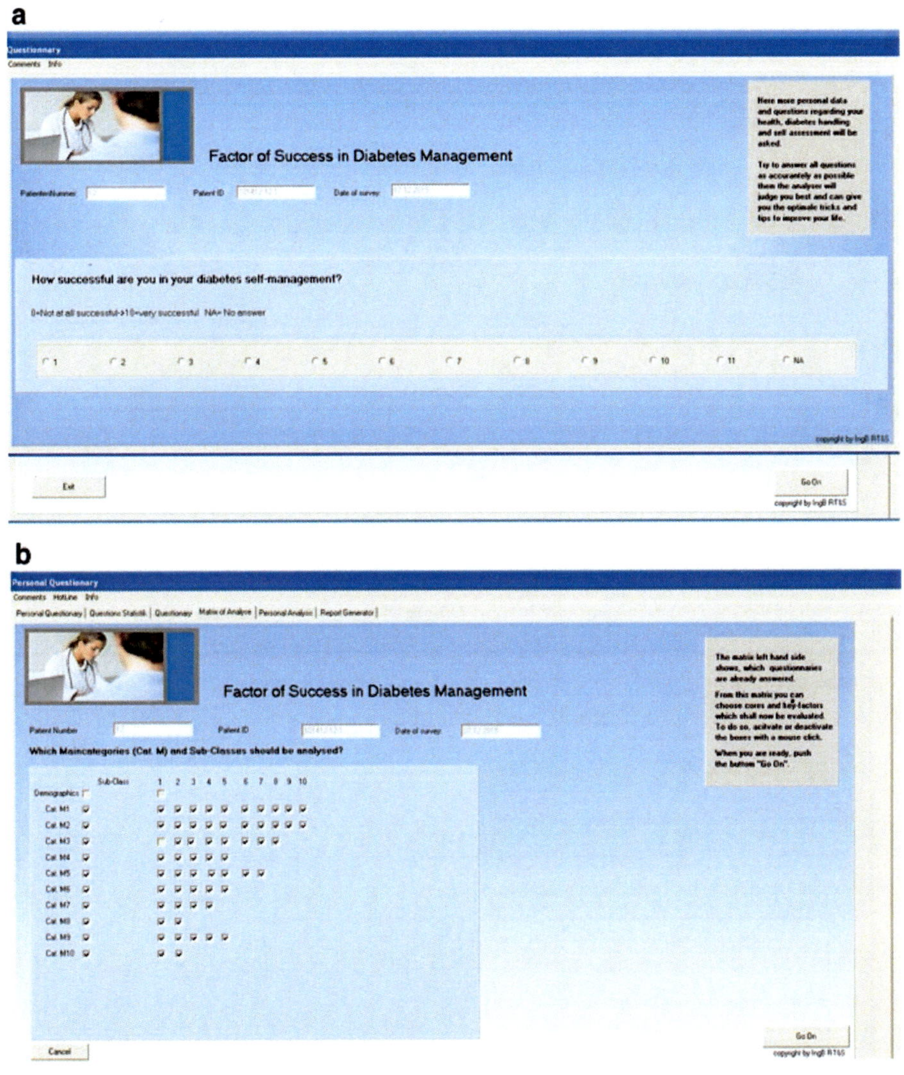

Fig. 8 **a** Question sheet of the questionnaire. **b** Decision mask of the questionnaire

three different trends, again on basis of our color coding and visualises them in special charts, as shown in Fig. 3. Out of these charts, another module calculates a trend of prognosis, in which statistical blips are taken into account. Furthermore, different time intervals for the trend of prognosis can be chosen, which gives the medical staff the opportunity to analyse momentary and slow changes. Moreover, the medical stuff or the patient themselves can choose alternate categorization of global status or the categorization by single questions. In that way, the tools identify the weak or the strong points in patient's behaviour and/or patient's handling of the disease. At least in that way, the doctors, the medical stuff, the social environment and the patient are empowered to supervise and interpret the momentary and past states of the patient's diabetes.

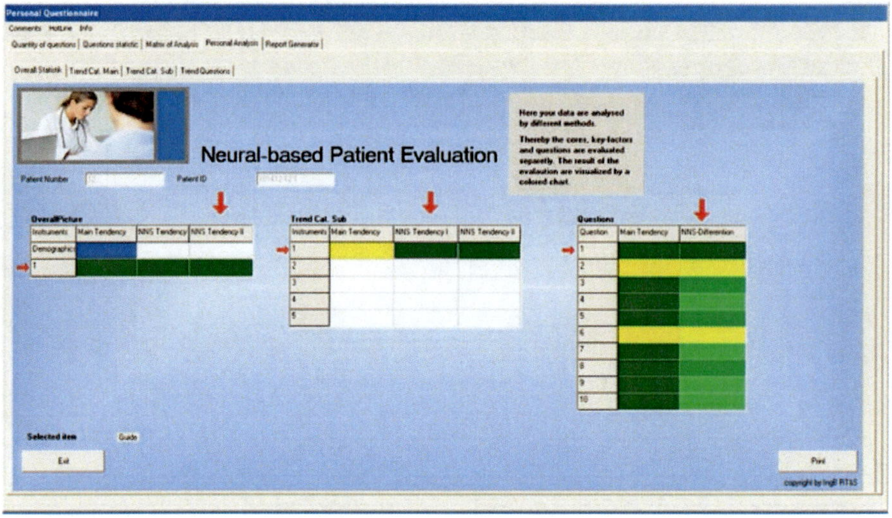

Fig. 9 Color-coded patient state analysis

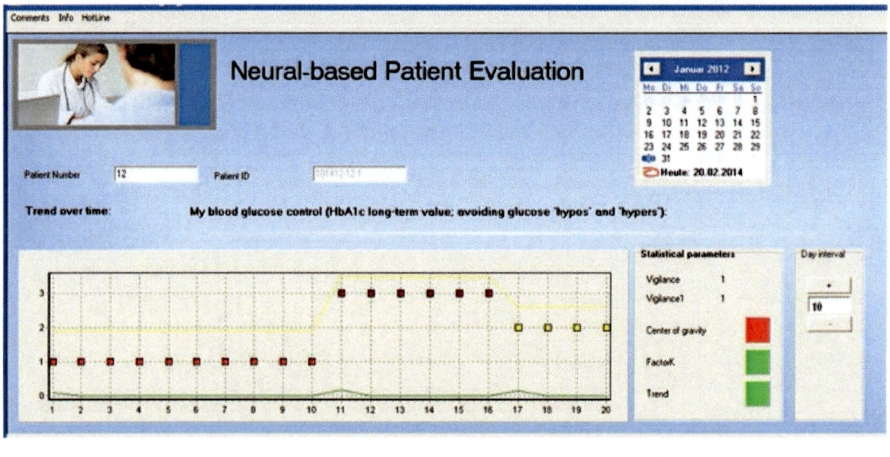

Fig. 10 Trend analysis and trend visualization of a patient over 20 examinations

Image 3 shows, in detail, the chart of the color-coded categorizations of the blood glucose control (HbA1c) of a patient over 20 examinations (left hand side). The values for the HbA1c (high or low) during the course of management can be easily pointed out by the Chart. The middle block shows the statistical parameters, the centre of gravity (the overall middle value over all examinations), and the Factor K, a special statistical process control which points out a prognosis for the next two weeks and the momentary trend over the next 10 days. Figure 10 shows a typical patient at the beginning of therapy. The HbA1c was high to low (red), but during the first therapeutic intervention, it shows improvement and thus the

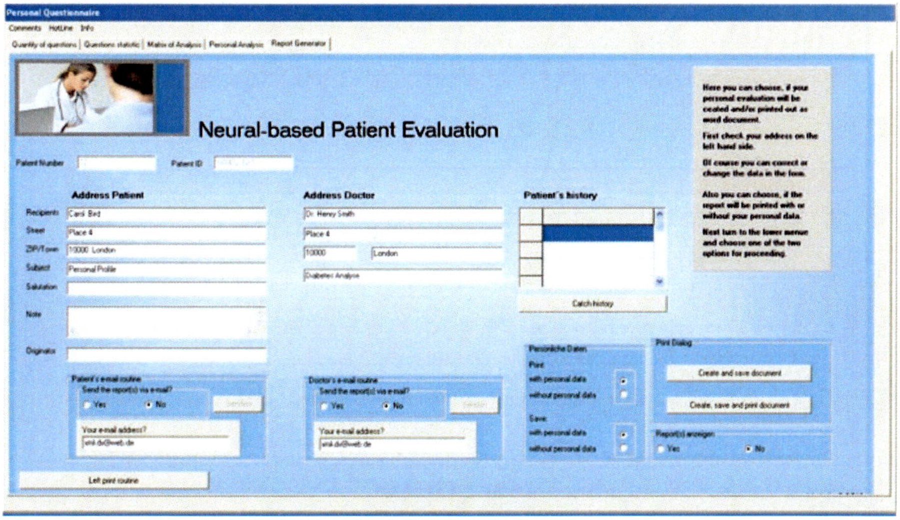

Fig. 11 User interface of the report generator

prognosis square shows a green color. As the duration of the improvement of the HbA1c value holds for a long time without significant fluctuation, the Factor K value square is green colored.

Our tool has the ability to archive the categorization and trend analysis in a PDF-document. This document can be stored in an anonymous or personalized form and can be printed out directly and/or sent via email to the patient or a medical centre. Furthermore, our report generator allows one to select further examination results for printing and/or controlling. Figure 11 shows the user interface of this reporting and printing dialogue with all optional possibilities of the documentation and sending routines.

A standard form of a report is shown in Fig. 12. At the top of the report, the personal data is (optional) given, followed by the legend of the color-coding. Next, the different questions' text and their classical and neural based categorization results are listed. The report of the questions ends with the chart of the completed questionnaires and the long and short time prognosis, coded again by the colors of the categorization and a short text segment.

4 Some Statistics of the Neural Net Condition Phase

We condition and test the system with two different data sets to explore the data sets of the partner company SMO Networks. The first data set, collected in the USA, contained 1024 patients' questionnaire data; 501 of Diabetes Type I and 523 of Diabetes Type II. The second data set, collected in Germany, contains 612 patients' questionnaire data of Diabetes Type I and 564 of Diabetes Type II. Different social,

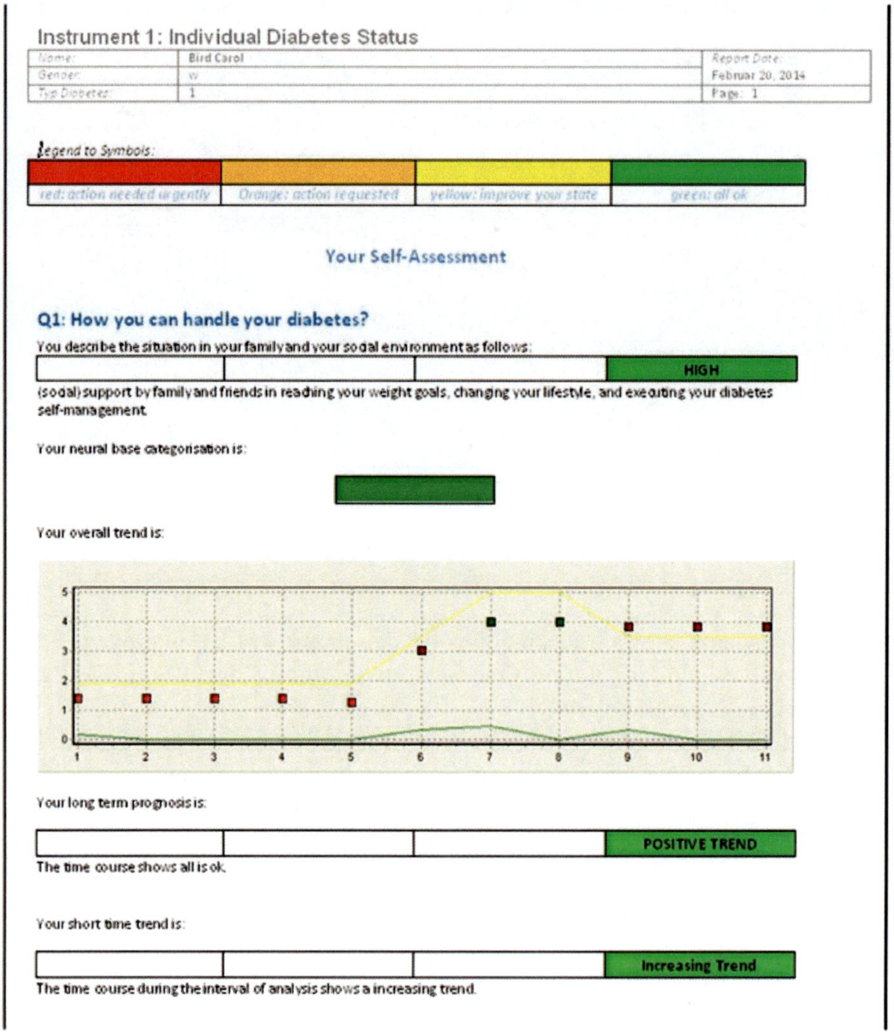

Fig. 12 Report of an exemplary patient categorization

ethnic and educational groups have been recorded to empower the neural nets to differentiate adequately between the wide ranges of data. After the training phase of the neural nets, the statistical distribution of the categorized data sets were evaluated and compared with the literature to ensure our system reflects reality. A typical statistical distribution of "bad data set categorization by a neural net" is shown by Fig. 13.

As can be clearly pointed out by Fig. 13, the neural net represents three of the four colors only (Column 1 represents the red color, column 2 the orange, column 3 the yellow and column 4 the green). However, it is important to remember that the reason for this may well lie in the unbalanced structure of the data itself, as well as in the conditioning of the neural network.

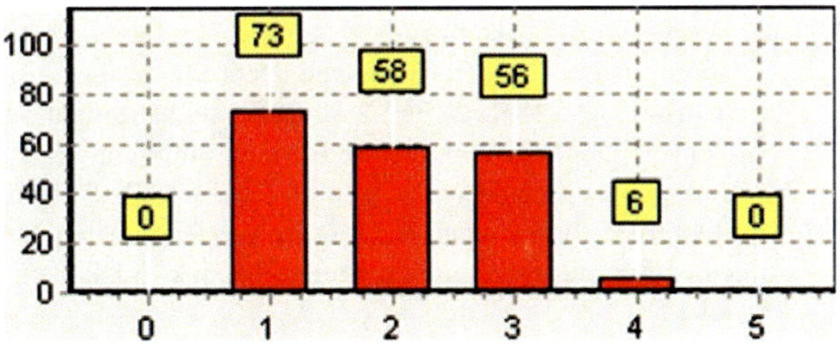

Fig. 13 Unbalanced statistical distribution of patient's data, not used for conditioning the neural net library

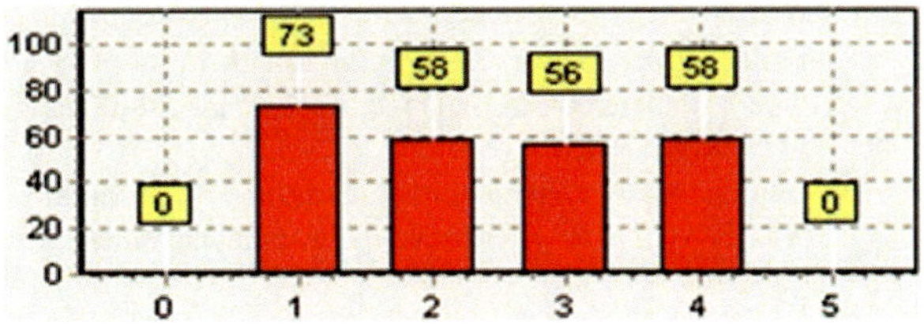

Fig. 14 Balanced statistical distribution of patients data, used for conditioning the neural net library

Figure 14 shows a well-conditioned net, based on a well-balanced data set. In this data set all four classes (all four color based categorization classes) are represented by nearly the same number of patients.

Using only such data sets, we conditioned a library of neural nets, representing all kinds of combinations of the questionnaires, where in each net all kinds of "possible patient's states" are stored.

Remark: To ensure the right categorization of extremely abnormal patients' data sets some simulated patient data sets with extreme states have to been taken to ensures that all kinds of patients can be categorized by the system in a logical and comprehensible way.

At least we evaluate the final system regarding its economy of time. For this we asked test persons to supervise and assess patients by common methods like paper or computer based patient medical files, Excel data sheets or conversation by comparison of using our system. After a short training phase, our test persons needed a factor ten times less for a patient evaluation; which means the system enables additional a fundamental cost saving.

References

1. American Diabetes Association, Diagnosis and Classification of Diabetes Mellitus. Diabetes Care **27**(1) (2004)
2. American Diabetes Association, Standards of medical care in diabetes—2014. Diabetes Care **37**(1), S14–S80 (2014)
3. Ellrodt, G., Cook, D.J., Lee, J., Cho, M., Hunt, D., Weingarten, S.: Evidence-based disease management. JAMA **278**, 1687–1692 (1997)

4. Szecsenyi, J., Rosemann, T., et al.: German diabetes disease management programs are appropriate for restructuring care according to the chronic care model: an evaluation with the patient assessment of chronic illness care instrument. Diabetes Care **31**(6), 1150–1154 (2008)
5. Reuter, M., Bohlmann, S.: Supervising multicut aggregates by special neural nets. In: World Automation Congress (WAC), 24–28 Juni 2012, Puerto Vallarta, Mexico (2012)
6. Reuter, M., Bohlmann, S.: Einsatz von neuronalen Netzen zur automatischen Überwachung einer Fertigungsanlage im Online-Modus. VDE-Kongress 2014 "Smart Cities". Frankfurt (2014)
7. Reuter, M., Bohlmann, S.: Personal diabetes management tools based on hybrid neural nets. In: BMT 2014—48th DGBMT annual conference, 8–10 October 2014, Hanover, Germany (2014)
8. Bailey, R.A., Pfeifer, M., Shillington, A.C., et al.: Effect of a patient decision aid (PDA) for type 2 diabetes on knowledge, decisional self-efficacy, and decisional conflict. BMC Health Serv. Res. (2015)
9. Garber, A., Abrahamson, M., Barzilay, J., Blonde, L., Bloomgarden, Z., Bush, M., et al.: AACE/ACE Comprehensive diabetes management algorithm 2015. Endo Pract. **21**, 438–447 (2015)
10. Nam, S., Chesla, C., Stotts, N.A., Kroon, L., Janson, S.L.: Barriers to diabetes management: patient and provider factors. Diabetes Res. Clin. Pract. **93**, 1–9 (2011)
11. King, D.K., Glasgow, R.E., Toobert, D.J., Strycker, L.A., Estabrooks, P.A., Osuna, D., et al.: Self-efficacy problem solving, and social-environmental support are associated with diabetes self-management behaviors. Diabetes Care **33**(4), 751–753 (2010)
12. Paget, L., Salzberg, C., Scholle, S.H.: Building a strategy to leverage health information technology to support patient and family engagement. http://www.ncqa.org/HEDISQualityMeasurement/Research/BuildingaStrategytoLeverageHealthInformationTechnology.aspx. Accessed 10 Oct 2015
13. Streja, D.A., Rabkin, S.W.: Factors associated with implementation of preventive care measures in patients with diabetes mellitus. Arch. Intern. Med. **159**, 294–302 (1999)
14. Paddock, E.P., Veloski, J., et al.: Development and initial validation of a patient satisfaction with pharmacy performance questionnaire (PSPP-Q) evaluation & the health professions **33** (2), 197–215 (2010)
15. Grogan, S., Conner, M., et al.: Validation of a questionnaire measuring patient satisfaction with general practitioner services. BMJ Qual. Saf. 9(4), 210–215 (2000)
16. Grogan, S., Conner, M., Willits, D., Norman, P.: Development of a questionnaire to measure patients' satisfaction with general practitioners' services. Br. J. Gen. Pract. **45**(399), 525–529 (1995)
17. Kohonen, T.: Self-organized formation of topologically correct feature maps. Biol. Cybern., 59–69 (1982)
18. McCulloch, W.S., Pitts, W.: A logical calculus of the ideas immanent in nervous activity. Bull. Math. Biophys. **5**, 115–133 (1943)
19. Rosenblatt, F.: The perceptron: a probabilistic model for information storage an organization in the brain. Psychol. Rev. **65**, 386–408 (1958)
20. Reuter, M., Bostelmann, S.: Computing with activities, WHU-structures and a quantization model of the neural nets, WSOM 2005, Paris, France (2005)
21. Reuter, M., Tadjine, H.: Computing with activities III: chunking and aspect integration of complex situations by a new kind of Kohonen Map with Whu-Structures (Whu-SOMs). IFSA. Beijing, VRC (2005)

Biological Inspired Image Analysis for Medical Applications

Matthias Reuter and Sabine Bohlmann

Abstract For the analysis of image data—independent of file format, recording mimic and image size—a general model of visual signal processing and interpretation was derived and converted into a bionic algorithm. This algorithm, which is based on the current neurobiological principles of the mammalian light sense, enables contrast enhancement and edge detection independent of illumination as well as implicit noise suppression without recourse to numerically standard image processing methods.

Keywords Image analysis · Bionic model · Neural networks · Computer vision

1 Introduction

The basic assumptions of this algorithm are based on two cybernetic model approaches for neuronal information processing and coding: On the one hand, the carrier-pattern meaning model of Benesch [1] and on the other hand the method of Computing with Activities of Reuter [2, 3]. What both approaches have in common is that a material carrier structure in the form of a neuronal net exists, into which information flows and evokes a meaningful activity pattern. Furthermore, only electrical and chemical potentials are available to nature for the processing, transmission and interpretation of information in order to trigger adequate reactions to incoming stimuli. These basic assumptions provide a theoretical basis for a non-numerical definition of signal analysis. The advantage of this type of coding, which is just based on potentials, is that its algorithmic description consists of only a few lower-level arithmetic operations and therefore provides very fast and robust results even for extensive or large image material [4, 5]. In the following parts of the paper, the basics and processing steps of the bionic algorithm are roughly described.

M. Reuter (✉) · S. Bohlmann
Department of Big Data and Technical Information Systems, TU Clausthal - Institute for Informatics/IngB RT&S, Clausthal-Zellerfeld, Germany
e-mail: matthias.reuter@tu-clausthal.de

M. Israelowitz et al. (eds.), *Biomimetics and Bionic Applications with Clinical Applications*, Series in BioEngineering,
https://doi.org/10.1007/978-3-319-53214-1_14

In order to understand these steps, the most important physiological basics of information processing in the retina of the mammalian eye will be shown. In the second part, the transformation of these neurophysiological basics into the bionic model are described and in the last sections some sample calculations will show the power of the bionic approach.

2 Physiological Basics of the Retina

A light stimulus that falls through the lens into the eye penetrates all existing cell layers and then hits the retina, where the conversion of the optical stimulus into an electrical signal by the photoreceptors begins. Signal transmission to downstream cells takes place via the synapses in the form of potential changes, i.e. a change in the released transmitters. In contrast to most other neurons, which code for stimulus strengths by means of action potentials, the strength of the stimulus in retinal neurons is derived from the extent of the voltage change at the cell membrane. Another important difference between the photoreceptors and common neurons is that an incoming stimulus (incidence of light) does not lead to depolarization and thus to an increased release of transmitters (glutamate), but hyperpolarizes the cell and reduces the number of transmitters. The downstream cells that further process the information due to the amount of glutamate released are the bipolar cells, which in turn transmit the information to the ganglion cells on a potential basis [6–9].

In order for adequate transfer of the information, there are two types of bipolar cells, which are distinguished by their excitation transmission and morphology. Based on the way they transmit excitation, they are subdivided into ON and OFF cells. As with photoreceptors, signal transmission takes place via graduated potentials, so that the intensity of a stimulus is encoded by the amount of emitted transmitters. Both the ON and OFF bipolar cells secrete the transmitter glutamate like the photoreceptors [10].

The decisive differences between the two bipolar cell types are their postsynaptic glutamate receptors, which react specifically to distributed transmitter amounts of the photoreceptors.

For example, receptors of ON bipolar cells react inhibitively to the neurotransmitter glutamate, which means that they do not release neurotransmitters in the dark and thus transmit little or no information. The OFF bipolar cell receptors, on the other hand, react to glutamate exitatorical, i.e. the cell is depolarised when glutamate arrives, so that in darkness it transmits information to the specific ganglion cells [6]. In the case of light incidence, the opposite is true: the ON bipolar cells are depolarized by the glutamate and the OFF bipolar cells are hyperpolarized [11].

The described information flow between the photoreceptors and the bipolar cells as well as between bipolar cells and ganglion cells is laterally influenced and thus modulated by interneurons—the horizontal and amakrin cells. The underlying neurotransmitters of the interneurons—mainly GABA—have an inhibitory effect on innervated cells, so that the literature often refers to lateral inhibition as a means of improving contrast perception [12, 13]. The following diagrams summarize the interaction of the different cell types (Figs. 1 and 2):

Without going into much detail, the described lateral influencing mechanism causes varying degrees of influence in the transition area between light and dark in a visual scenery [15, 16]. The final response behaviour clearly shows the contrast enhancement at the transition between the two brightness levels. The value in the dark transition area is less strongly reduced by the lower inhibitory contribution from the light area compared to the uniformly dark areas, while the value in the light area is more strongly reduced by the high inhibitory contribution from the dark area than the regions with constant brightness. Thus, the following potential curve can generally be recognized at light-dark transitions (Fig. 3).

In order to further describe the lateral influence of ON- and OFF-paths, it is essential to consider the arrangement of receptors and neurons. The retina relies on so-called receptive fields, which are based on a similar antagonistic mode of action and the separation of information about ON and OFF paths as with bipolar and ganglion cells. Receptive fields are divided into ON-Centre-OFF-Surround and OFF-Centre-ON-Surround [17, 18]. Figure 4 on the left shows a small receptive field whose centre consists of a photoreceptor, while on the right large receptive fields are shown in which several photoreceptors always form the centre of a receptive field. The respective photoreceptors in the surroundings form the surround sound.

Since the antagonistic surround areas are caused by the inverted synapses of the horizontal cell, the bipolar cells in these ON-Centre-OFF-Surround receptive fields respond with a maximum depolarization at an incidence of light that fills the centre precisely, while the surround area is not illuminated. The associated ganglion cells

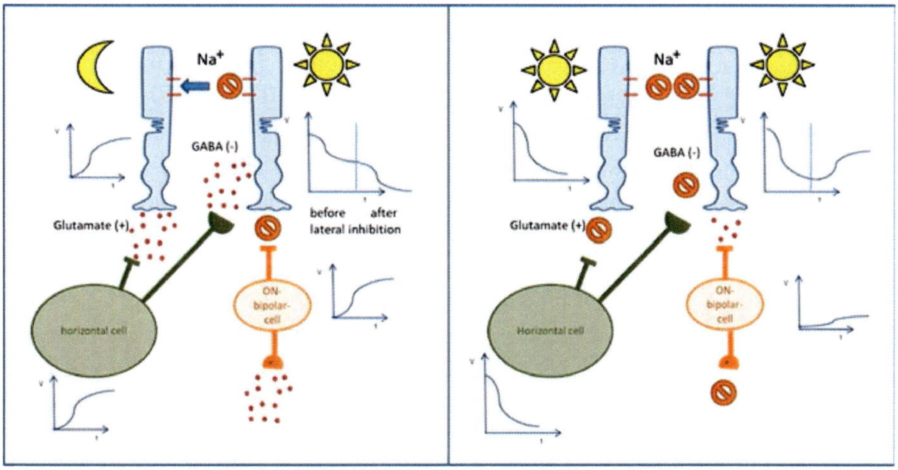

Fig. 1 Information path from the photoreceptor to the ON bipolar cell, left: Centre brighter than surround, right: same illumination intensity in centre and surround [14]

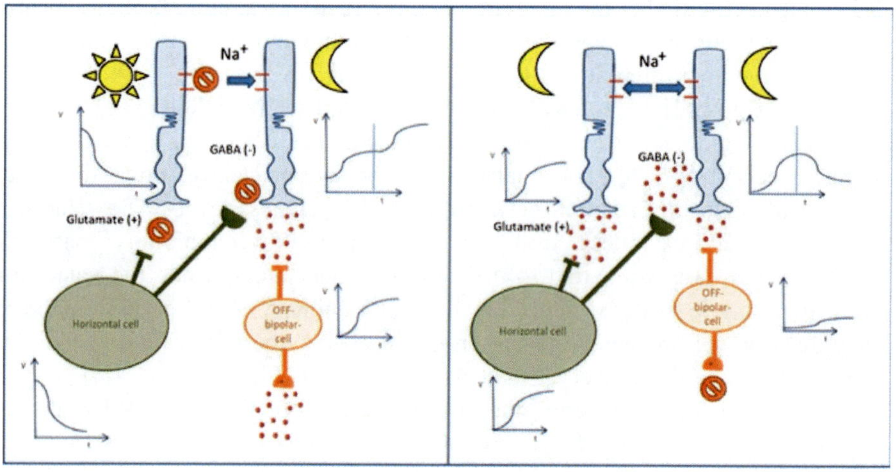

Fig. 2 Information path from the photoreceptor to the OFF bipolar cell, left: Centre brighter than Surround, right: same illumination intensity in centre and Surround [14]

Fig. 3 Potential curve at light-dark transitions [14]

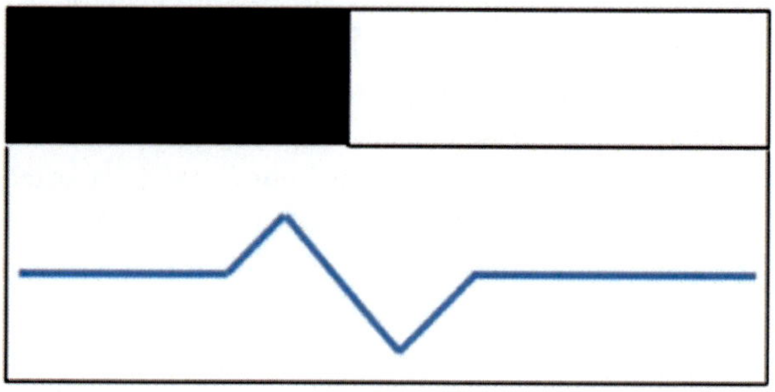

then fire with a high frequency of action potentials. The depolarisation or the sequence of action potentials is weaker the less light that hits the centre of the receptive field. If the light in this type of receptive field strikes both the centre and the surroundings, the associated bipolar cell only reacts with its basic activity, so that the threshold potential of the ganglion cell is not reached and it therefore does not trigger any action potential. An exclusive stimulation of the surround area causes a hyperpolarization of the bipolar cell, as a result of which no action potentials are generated by the ganglion cell. The strength of the hyperpolarization also depends on how much the surround area is illuminated [17].

With OFF-Centre-ON-Surround receptive fields, the reactions described above proceed in an opposite manner to ON-Centre-OFF-Surround receptive fields. For example, OFF-Centre bipolar cells react with maximum depolarization when the centre is not illuminated while the surround area is fully illuminated. The associated ganglion cell then fires a high frequency of action potentials.

This ON/OFF organizational structure ensures that changes in bright scenes can be detected just as well as in dark ones, i.e. the actual illumination of a visual scene

Fig. 4 Types of receptive fields, above: ON-Centre-OFF-Surround receptive fields, below: OFF Centre ON Surround Receptive fields [17]

is insignificant, only the gradient between the differences in brightness is relevant. This way, nature makes it possible to detect edges and contrasts independently of illumination.

The strictly biological and physiological basics in the first part form the basic parameters for the bionic model of image processing, which implements contrast enhancement and implicit noise suppression. Changes in the structure of images that are represented by different values of the pixel grey and/or colour values can be equated with neighbouring photoreceptors of different illumination intensities.

Thus, if the structure of an image changes, it manifests itself through increased activity of the ON or OFF bipolar cells. The changed illumination gradient leads to a change in the receptor potential and thus to a change in the activity pattern. If there is no change in an image—i.e. if I have a homogeneous surface—there is no change in the receptor potential and thus in the activity pattern, which is expressed by the fact that neither the ON nor the OFF bipolar cell reacts to this homogeneous stimulus.

Of decisive importance in this process is therefore the lateral influence, which enables the recognition of structural changes due to altered transmitter release and the resulting separation of the information pathways into ON and OFF paths.

3 Bionic Model for Image Analysis

The model description of the influence of the horizontal cell on the centre pixel is based on the physiology of the photoreceptor, which is divided into three segments: outer segment, axon, and cell membrane. These segments are designed in the bionic model of information processing as follows:

1. Light input ($\triangleq$ outer segment) = Segment 1
2. Segment receiving lateral influences ($\triangleq$ Axon) = Segment 2
3. Potential forming cell membrane = Segment 3.

Segment 1 takes the incident intensity of the light falling on the photoreceptor. Transferred to the model, this can be equated in an image with the predominant pixel value at any point in the image, which at the same time forms the centre of a receptive field.

Segment 2, which receives the lateral influences from the horizontal cells, corresponds to the potential of the horizontal cell in the model, i.e. the pixel values of the neighbouring pixels located in the surround of the centre pixel.

Segment 3 calculates these two segments into a modified photoreceptor potential and a resulting transmitter distribution.

By subdividing the photoreceptor into separate segments, it is possible to show that each segment initially forms its own independent potential—light induced at one point and laterally influenced at the other. Thus, there is no simple summation or addition of the potentials, but an integration—i.e. a recomposition of different influencing factors—to a modified potential, which reflects the new value of the deviation of the photoreceptor from its rest state (Segment 3). In this way, a new modified input value is generated from both potentials, which acts on the photoreceptor (Fig. 5).

By calculating a new input value of the photoreceptor, which hyperpolarizes it from its resting potential, one loses the actual information about the intensity of the light intensity acting on the photoreceptor at this point in the model. The new potential no longer contains only the information about the initial pixel value, but also the measure of the influence of the lateral inhibition acting on the centre pixel. The calculated input value is therefore a measure of how strongly the centre pixel is hyperpolarized. The resulting potential, on the other hand, is a measure of the difference between centre and surround.

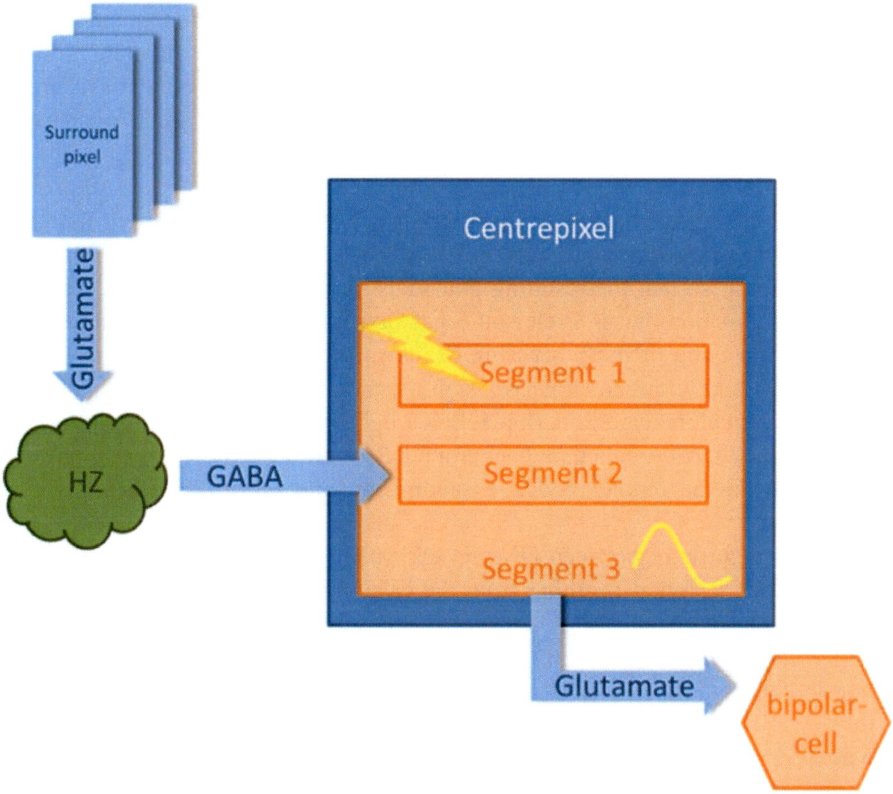

Fig. 5 3-segment model of the photoreceptor [14]

In concrete terms, this means that if, due to a low potential value (bright pixels) from the environment, a low lateral inhibition acts on a dark centre pixel, this has the consequence that the already low input value (pixel value) of the centre pixel is changed only slightly and thus the photoreceptor is hyperpolarized only slightly. The potential and the associated transmitter output of the photoreceptor are therefore correspondingly high and can be interpreted as a sign that the gradient between the centre and surround light intensities is large. This means that there is a significant change in the structure of the image in the transition between the centre and surround areas, i.e. there is a high probability of an object boundary. The darker the surround pixels become—the more similar they become to the centre pixels—the stronger is the lateral influence of the horizontal cells, since more GABA is emitted. This means that the contribution by which the photoreceptor in the centre is brought out of its resting potential also increases. The resulting transmitter output is correspondingly lower and is a sign that there are structural changes in the image, but that these changes are not as pronounced as in the first case. If the inhibitory contribution of the horizontal cells is approximately equal to the potential of the centre pixel, a new input is obtained that does not announce any serious changes between the light intensities of the centre and the surround. The amount of emitted transmitters on the level of the bipolar cells is now an indication for a uniform input, which means that no significant transitions—i.e. object boundaries—are to be expected at this point of the visual scenery. If we look at a bright pixel with its surroundings, the same effects of lateral influence occur.

Common features when viewing both the dark and the bright centre pixels are that small differences between the light intensities in the centre and surround due to the antagonistic centre and surround areas balance the inputs flowing into the photoreceptor, so that the resulting potentials for information processing represent diffuse information that is irrelevant for the recognition of object boundaries.

The main thing of the processing of information in ON and OFF paths is that both a high and a low potential is a sign of a strong contrast in the visual scenery. A high potential indicates a dark pixel on a light background, while a low potential indicates a light pixel on a dark background. So, in order to process this information adequately in the bionic model, a subdivision into ON and OFF paths is also necessary.

By distinguishing between ON and OFF paths, it is possible that the low neurotransmitter secretion resulting from the calculation is not synonymous with less information being passed on to the cortex, since the bipolar cells downstream of the photoreceptors react differently to the presence or absence of the transmitter glutamate. OFF bipolar cells form excitatory synapses with the photoreceptors so that they depolarize in darkness, ON bipolar cells have inhibitory synaptic connections with the photoreceptors so that they hyperpolarize in darkness and depolarize in brightness. This means that due to darkness a high amount of transmitter is released from the photoreceptor, which in turn triggers a high potential at the OFF bipolar cell, while the absence of glutamate at brightness results in a high potential at the ON bipolar cell. For a medium intensity pixel value, this does not mean that a medium intensity potential is triggered and passed on, but that only a weak one in either the ON or OFF path, depending on whether it is above or below the cut-off point. This corresponds to the behavior of the neurons in the retina, which react only with a basic activity with uniform light intensity in the receptive field.

This described transformation process from the modified input value to a potential of the ON and OFF bipolar cells is illustrated in Figs. 6 and 7.

4 Sample Calculation

The left illustration shows an image with two homogeneously illuminated areas, the left side being dimly illuminated and the right side strongly illuminated, so that—as shown below—a high potential is generated on the left side and a low potential is generated on the right side. Due to the lateral influence of the surround to the centre, the areas without contrast are normalized to a medium potential, so that they do not represent any relevant information for the bipolar cells and they therefore only react to the input with their basic activity. At the transitions, however, the contrast is

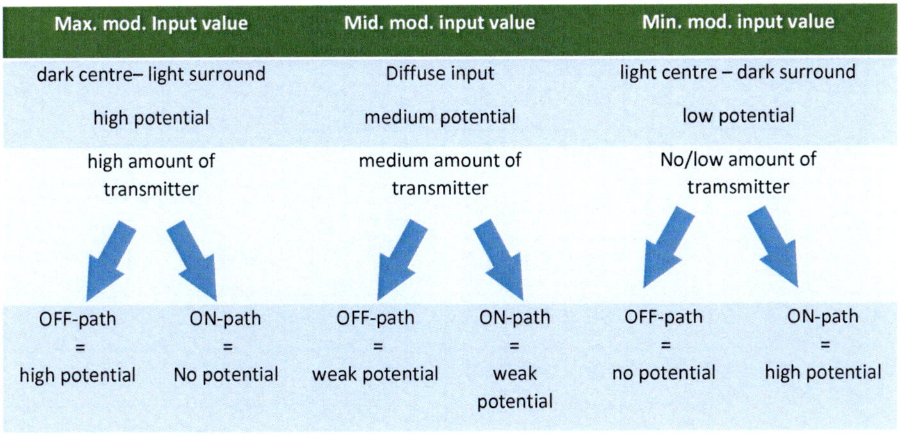

Fig. 6 Conversion of the input values into potentials for the ON and OFF path [14]

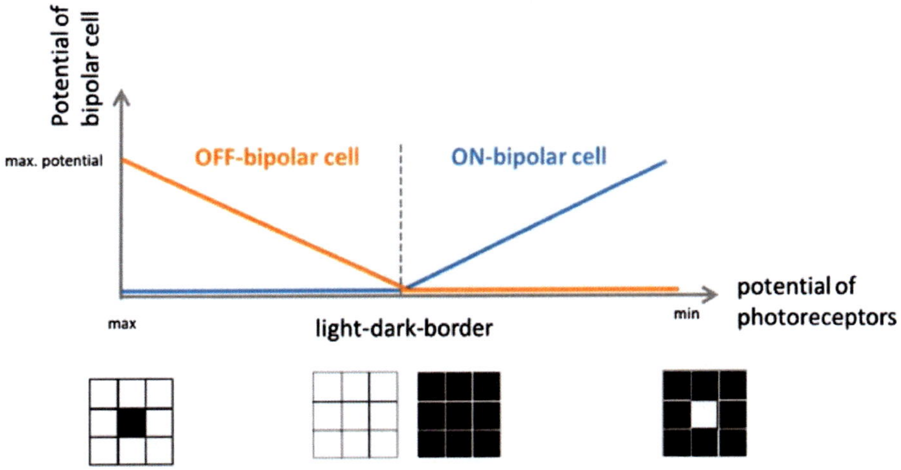

Fig. 7 Transformation of modified photoreceptor potentials into potentials of ON- and OFF-bipolar cells [14]

emphasized by generating a higher and a lower potential than the basic activity. If these potentials of the photoreceptors are now transferred to the respective bipolar cells, it can be seen that the OFF and ON bipolar cells are excited once and react to the input with an equally high potential (Fig. 8).

The same applies to individual lines or other structures. The following example shows that edge detection can be realized even with minimal differences in the pixel values due to the unique separation in ON and OFF paths. Figure 9 shows a vertical bar with differences in the pixel values of 1 (the bar has the pixel value 33 and the surround 32).

10	10	10	80	80	80	80
10	10	10	80	80	80	80
10	10	10	80	80	80	80
10	10	10	80	80	80	80
10	10	10	80	80	80	80
10	10	10	80	80	80	80
10	10	10	80	80	80	80

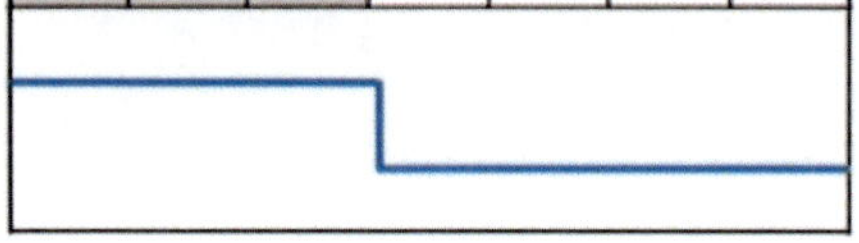

50	50	63,12	36,87	50	50	50
50	50	63,12	36,87	50	50	50
50	50	63,12	36,87	50	50	50
50	50	63,12	36,87	50	50	50
50	50	63,12	36,87	50	50	50
50	50	63,12	36,87	50	50	50
50	50	63,12	36,87	50	50	50

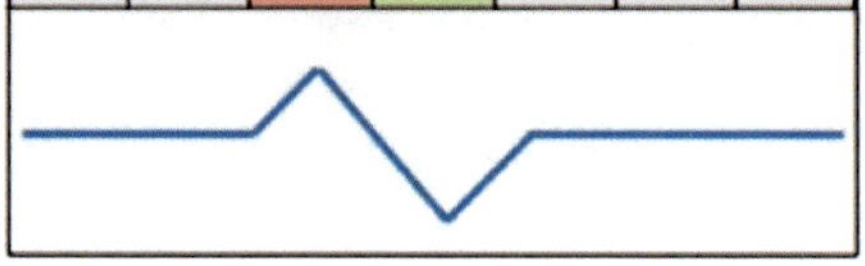

Fig. 8 Left: Pixel value image due to the incident light, right: potential image of the gradients of centre and surround, bottom: the potential curve in each case [14]

Fig. 9 Image of modified photoreceptor potentials and color marking of ON- and OFF-bipolar cell activity [14]

50	50	49,94	49,94	49,94	50	50
50	50	49,88	50,44	49,88	50	50
50	50	49,81	50,38	49,81	50	50
50	50	49,81	50,38	49,81	50	50
50	50	49,81	50,38	49,81	50	50
50	50	49,88	50,44	49,88	50	50
50	50	49,94	49,94	49,94	50	50

Based on the potentials of the photoreceptors represented in numbers and the colored marking of the activity of ON and OFF bipolar cells, it becomes clear that even with small differences in light intensity, an automatic subdivision into ON and OFF paths takes place. This means that the developed model is very well suited for the detection of smallest contrast transitions in a visual scenery, as each change is announced by the separation into ON and OFF paths. The absolute light and dark intensities are irrelevant, only the gradient between the observed centre and the corresponding surround is decisive.

5 Medical Applications of the Bionic Model of Image Analysis

Figure 10 shows the section through the hypothalamus of a mouse brain and Fig. 11 the intermediate filament in brain tissue. The images were analysed using the bionic filter to differentiate between individual cell types.

Both images show that the outlines of the individual cells can be visualized very well, i.e. the algorithm has captured the contours and separated them correctly.

In general, it can be said that the model and its algorithmic implementation make it possible to work out tissue structures and separate different tissue compositions. It is irrelevant what type of tissue is involved and what task is pursued.

In applying these results to larger structures and tissues and to evaluate the general applicability of the bionic filter, two further example images are analysed below. Figure 12 top left shows a thorax image in which the lung and bronchial carcinoma (arrow) are visualized by volume rendering [21]. Figure 13 left shows the X-ray image of a forearm fracture [22]. The analysis of these images is intended to show that it is possible to highlight/separate the carcinoma and the captured

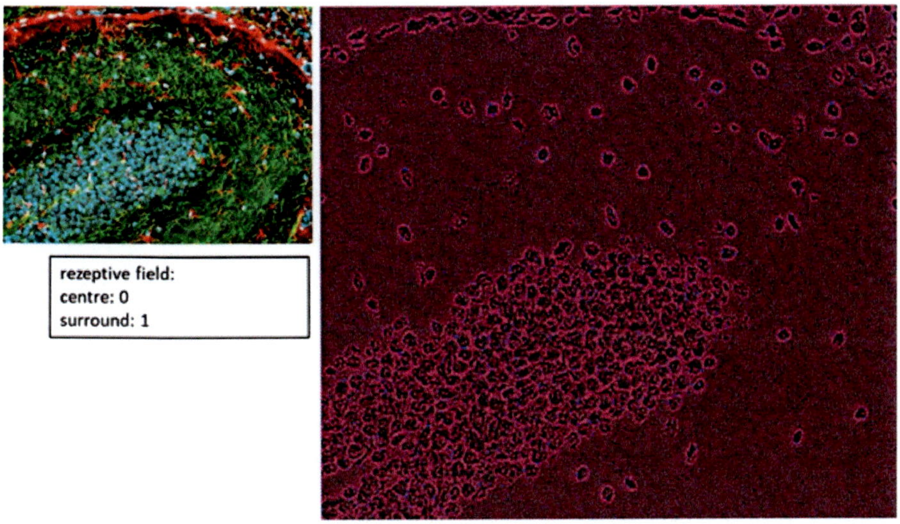

Fig. 10 Example "Hypothalamus" [19], left: Original test image and settings of the bionic filter, right: evaluation by the bionic filter, basic setting when selecting the receptive field 3 × 3 surround pixels, one centre pixel

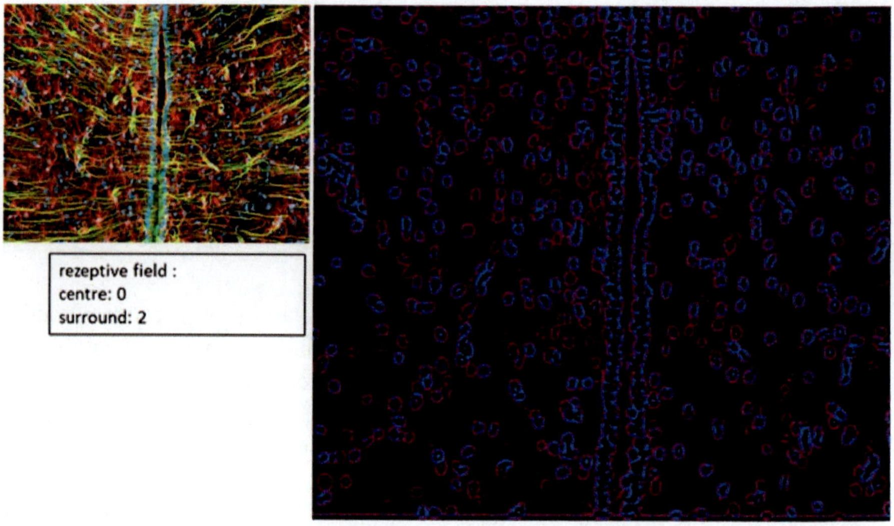

Fig. 11 Example "Filament" [20], left: Original test image and settings of the bionic filter, on the right: evaluation using the bionic filter, basic setting when selecting the receptive field 7×7 surround pixels, one centre pixel

organs as well as to highlight/separate bones together with the tissue surrounding them and a bone structure.

In all three result images from Fig. 12, the carcinoma can be clearly separated from the surrounding tissue. The selection of a small receptive field (top right) allows an exact tracing of the carcinoma outline, while the selection of large receptive fields (bottom right) also leads to a better separation of the individual organs shown in the picture.

The presentation of the evaluation result from the test image "Fracture" shows very clearly that the selection of a large receptive field (centre) enables a 3-dimensional representation of the bones, which emphasises the fracture point significantly better. Compared to the original image, the blurred impression is converted into a clearly structured representation so that tissue and bone can be clearly separated from each other. A similar effect is achieved by selecting an asymmetric receptive field (Fig. 13 right).

It can therefore be seen that the bionic filter is not only capable of working out tissue structures, but also larger areas such as organs, tumours, bones and associated fractures.

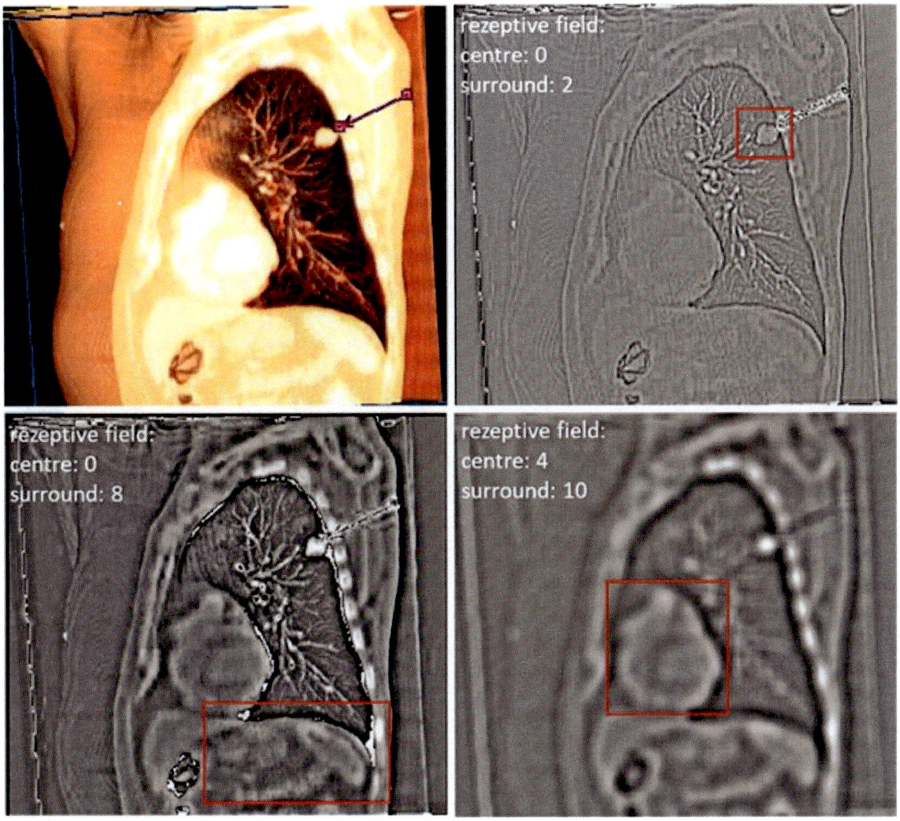

Fig. 12 Test pattern "Thorax", top left: Original [21], top right: Result of the analysis using the bionic filter, basic setting when selecting the receptive field 7 × 7 surround pixels, one centre pixel, bottom left: Result of the analysis using the bionic filter, basic setting when selecting the 19 × 19 surround pixel receptive field, one centre pixel, bottom right: Result of the analysis using the bionic filter, basic setting when selecting the receptive field 23 × 23 surround pixels, 9 × 9 centre pixels

Using this new approach, a variety of medical image processing tasks such as organ segmentation to determine the physical size or tumour segmentation to define the target region and plan treatments could be performed. The new model and its algorithm thus represent an effective approach to eliminating the shortcomings of classical methods—namely the choice of analysis method depending on the objective and task—and to develop a general method for medical image processing.

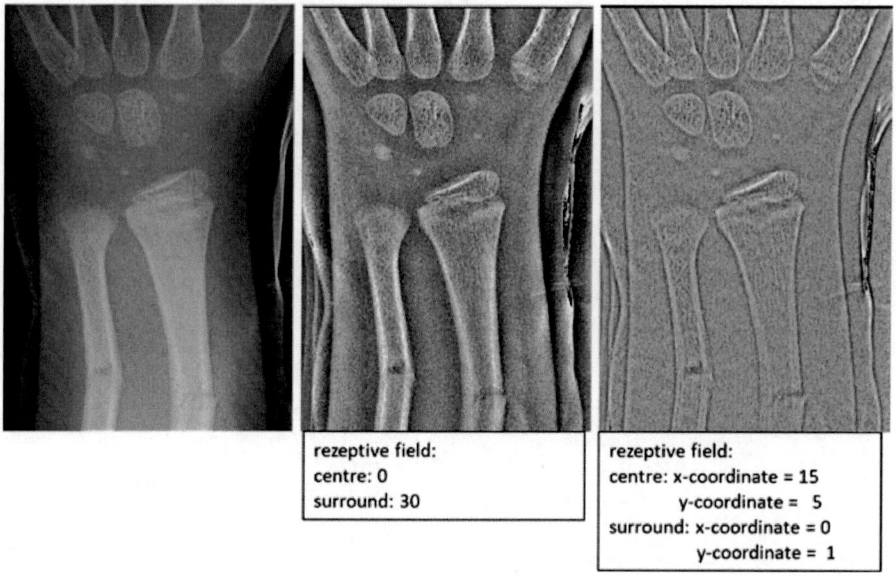

Fig. 13 Test pattern "Fracture", left Original [22], centre: Result of the analysis by the bionic filter, basic setting when selecting the receptive field 63 × 63 surround pixels, one centre pixel, right: Result of the analysis by the bionic filter, basic setting when selecting an asymmetric receptive field 33 × 13 surround pixels, 1 × 3 centre pixels

References

1. Benesch, H.: Der Ursprung des Geistes. München: dtv 1542 (1980)
2. Reuter, M., Tadjine, H.: Computing with activities III: chunking and aspect integration of complex situations by a new kind of Kohonen Map with Whu-Structures (Whu-SOMs). IFSA. Beijing, VRC (2005)
3. Reuter, M., Bostelmann, S.: Computing with activities, WHU-structures and a quantisation model of the neural nets. WSOM. Paris, France (2005)
4. Reuter, M., Bohlmann, S.: Analysing radargrams and sonargrams by biological inspired signal processing methods to optimize the detection of mines and dumped ammunition. In: 10th security research conference, future security 2015, 15–17 September 2015, Berlin, Germany (2015)
5. Reuter, M., Bohlmann, S.: About new—biological inspired—pattern analysis methods for the pattern recognition in radargrams. In: International radar symposium IRS-2015, 24–26 Juni 2015, Dresden, Germany (2015)
6. Masland, R.H.: The neuonral organization of the retina. Neuron **76**, 266–280 (2012)
7. Goldstein, E.B.: Wahrnehmungspsychologie. Spektrum Verlag, Heidelberg (2002)
8. Koester, J., Siegelbaum, A.: Membrane potential and the passive electrical properties of the neuron. In: von Kandel, E.R., Schwartz, J.H., Jessell, T.M., Siegelbaum, S.A., Hudspeth, A. J. (eds.) Principles of neural science, pp. 126–147. McGraw-Hill Companies (2013)
9. Koester, J., Siegelbaum, A.: Propagated signaling: the action potential. In: von Kandel, E.R., Schwartz, J.H., Jessell, T.M., Siegelbaum, S.A., Hudspeth, A.J. (eds.) Principles of neural science, pp. 148–171. McGraw-Hill Companies (2013)
10. Euler, T., et al.: Eyecup scope—optical recordings of light stimulus-evoked fluorescence signals in the retina. Pflugers Arch, 1393–1414 (2009)

11. Campbell, N.A., Reece, J.B., Markl, J.: Biologie. Spektrum Akademischer Verlag (2003)
12. Morgan, I.G.: What do Amacrine cells do? Prog. Retin. Res., 193–214 (1991)
13. Boycott, B.B., Wässle, H.: The morphological types of ganglion cells of the domestic cat's retina. J. Physiol., 297–419 (1974)
14. Bohlmann, S.: Neuer Ansatz für die Vorverarbeitung von Bilddaten mittels Umsetzung der retinalen Informationsweitergabe in ein Potential-orientiertes Neuronen-Modell, Diss., TU Clausthal (2016)
15. Gollisch, T., Meister, M.: Eye smarter than scientists believed: neural computations in circuits of the retina. Neuron **65**, 150–164 (2009)
16. Mallot, H.A.: Computational vision information processing in perception and visual behaviour. The MIT Press, Cambridge u.a (2000)
17. Hubel, D.H.: Eye, brain, and vision. (dt.: Auge und Gehirn. Neurologie des Sehens). San Francisco (dt.: Heidelberg), (dt.: Spektrum Akademischer Verlag), 1995 (dt.: 1989)
18. van Wyk, M., Taylor, W.R., Wässle, H.: Receptive-field properties of ON- and OFF-ganglion cells in the mouse retina. Vis. Neurosci., 293–308 (2009)
19. Molecular Expressions. http://micro.magnet.fsu.edu/micro/gallery/brain/brain3large.jpg (Zugriff am 8. August 2015)
20. Molecular Expressions. http://micro.magnet.fsu.edu/micro/gallery/brain/brain4large.jpg (Zugriff am 8. August 2015)
21. Wikimedia Commons. https://commons.wikimedia.org/wiki/File%3AThorax_CT_pcor_Cut_3D_Volume_Rendering_Haut.jpg (Zugriff am 13. Oktober 2015)
22. Wikimedia Commons. https://commons.wikimedia.org/wiki/File%3AGruenholzfraktur_-_Unterarmfraktur.jpg (Zugriff am 13. Oktober 2015)

Author Index

B
Bohlmann, Sabine, 81, 147, 159, 177, 195, 211

F
Fließ, Malte, 61

G
Gifford, Jenifer, 71
Gille, Christoph, 47, 81, 147, 177

I
Israelowitz, Meir, 47, 81, 147, 177

K
Kramer, James, 177
Kull, Kalevi, 3

L
Leming, Matthew, 71

M
Michael, Obaro S., 19, 131
Müller, Robert, 47

N
Nöhre, Mariel, 107

R
Reimers, Kerstein, 47, 81, 147
Reuter, Matthias, 81, 147, 159, 177, 195, 211
Rizvi, Syed W.H., 47, 81, 147, 177

S
Schmälzlin, Elmar, 107
Schüler, Dirk, 47
Strauß, Sarah, 47, 61

T
Thomas, Henkel, 47

U
Uebe, René, 47

V
Vogt, Peter M., 29, 47, 81, 147
von Schroeder, Herbert P., 47, 81, 147, 177

W
Wagner, Robert, 81
Weyand, Birgit, 29, 47, 81, 107, 147, 177

M. Israelowitz et al. (eds.), *Biomimetics and Bionic Applications with Clinical Applications*, Series in BioEngineering,
https://doi.org/10.1007/978-3-319-53214-1